CONTENTS

Sl. No.	Topic	Page No.
1.	Remembering Dr. A. P. J. Abdul Kalam He was a true Karmayogi -People's President	1
2.	Knowledge Sheet for becoming a Space Scientist	7
3.	Scope for Space Research in the world	11
4.	Rocket and Equations-School Level rocketry	13
5.	Rocket Engines used in the world	27
6.	Satellite and Orbits	32
7.	Global Positioning System (GPS)	40
8.	International Space Stations (ISS)	44
9.	Global Space Industry (GSI)	51
10.	International Space based- Solar Power Stations (ISBSP)	53
11.	Future Power without wire from infinite space	55
12.	Desalination of Sea water for potable water	56
13.	Summary of Solar Power Generation in Space	61
14.	Mission to Moon and Mars	68
15.	Perspective and Challenges in Mars Mission	70
16.	Reusable Launch Vehicle for cost reduction	73
17.	Global Warming and Climate Change	75
18.	Green House Effect	77
19.	Human Stress Management	82
20.	ആഗോള താപനവും കാലാവസ്ഥാ വ്യതിയാനവും വെല്ലുവിളികളും	94

[References-The articles on space programmes of different countries]

Book

English
**INFINITE SPACE &
UNLIMITED EXCITEMENT**

June 2018

Written & Published by
**Er. B Gopakumar
Former Senior Scientist, VSSC, ISRO,
Thiruvananthapuram, kerala**
House No. TENRA.54, Vaisakh, Thycaud
Elankam Nagar C Lane,
Thycaud P.O., Trivandrum - 695 014,
Kerala Mob : 9446550034
E-mail: bgopakumar21835@gmail.com

Copies:
50

Price
₹ 299/-

Layout & Design
Vijaya Press, Vazhuthacaud,
Trivandrum-695 014

© **All rights reserved**. No part of this book may be reproduced or utilized in any form or by any means, electronic or digital including photocopying or by any information storage or retrieval systems without the permission in writing from the author/publisher.

PREFACE

Respected readers,

It gives me great pleasure to publish this book "infinite space and unlimited excitement" related to space technology for the benefit of common man. The objective of this knowledge sheets are particularly aimed at the evolution through knowledge targeted to the higher secondary school, engineering college level students and layman.

When students study the basics in class rooms they may not be able to imagine the numerous future applications for developing technology and importance of studying fundamentals of Physics, Chemistry, Mathematics, English and various engineering subjects etc. Teachers must have a major role in this regard to generate interest in students to study all the fundamentals scrupulously.

India is blessed with youth power. Unfortunately this power is underutilized. The rulers of this country must give importance in utilizing this unique youth power for developing agricultural production, energy management, drinking water production, cost effective health care, pollution control, waste management etc through value added education. Employment opportunity to all must be an important requirement for the people to live in India with prosperity.

All development areas required large scale needs of advanced technology with cost effective practical inputs. Space technology also has a major role for creating opportunities to overcome the challenges and perspective in made in India products through Space industry.

This knowledge sheets may throw light on the possible application of the subject they study in school and college and illuminate their thinking process for future innovations particularly in global space industry. As you are aware, the inspiring innovations of space technology and infinite space and unlimited excitement are not that very simple to explain through a small book like this. As per the central Government guidelines for superannuation of employees they have to share their knowledge in Google pensioner's portal- Anubhav. Since I also superannuated from central Government service as senior scientist from Vikram Sarabhai Space Centre, ISRO, Trivandrum, would like to share my little knowledge with readers.

A candle loses nothing by lighting another candle. I hope readers will find it so on academic interest. While reading this book you may feel that some chapters are like science fiction or fantasy. I wish that fictions will come to reality in the near future. The incredible journey of Dr Elon Musk a living legend in rocketry in USA is a role model to space industry. He is the, owner of Space X private company, and supplier of Falcon Heavy reusable rocket to NASA and dreaming human habitat in Mars. Yesterday's dream; tomorrow's reality.

History proves that technology is the driver for development and civilization of mankind. It could be used as well as misused. The attitude of the user had much to do with its applications. My inspiration comes from seeing the enthusiasm of youth community, contributing to achieve many milestones for the development of India.

I strongly believe that we Indians can rule the technology domain for generations to come, making this planet more livable with love, peace and prosperity. This book is dedicated to my wife Radha Gopan, daughters Karthika Anuraj, Gowri Jithin, granddaughter Anika and grandson Saketh who support me to live with happiness.

Thank you

Er. B Gopakumar
Former Senior Scientist. VSSC , ISRO. Trivandrum.
House No.TENRA.54.VAISAKH.Thycaud Elankam Nagar.
Thycaud P.O., Trivandrum- 695014, Kerala Mob.9446550034.

REMEMBERING Dr. A P J ABDUL KALAM (1931-2015)
HE WAS A TRUE KARMA YOGI- PEOPLE'S PRESIDENT

Writing a tribute to Dr.Kalam is rather a challenging task. All Indians remember him as a person who is extremely knowledgeable. He proved himself that knowledge is power. Sharing his knowledge with others made him powerful. He is known as missile man of India. It is difficult to keep Dr Kalam away from children and children from him. He loved simplicity and hate ego.

When I joined Vikram Sarabhai Space Centre, Thiruvanathapuram, he was the project Director of SLV 3 rocket project. The first powerful rocket made in India helped the country to become a leader in space technology, under his leader ship. The knowledge he gained from ISRO had shared with DRDO and developed powerful missiles made in India. These achievements had given a thrust to his growth to get Bharatha Ratna award. Finally he became the President of India, the first bachelor and scientist at Rashtrapati Bhavan. The passion to share knowledge was his hallmark. He honored every chair that he held, and India honored him with the best it could and acclaimed as the peoples president for his warmth and accessibility. .

He was a space scientist first, the President and everything else later. Hence I would like to introduce him as an ISRO scientist worked in Thumba beach for more than 20 years. VSSC colleges called him as Veda Vyasen of 21 century. For them he was a Guru ,who asked to dream to fly and explore beyond horizon. A team player with a mission to ignite the minds of India's younger generation by the quote **"youth below the age of 25 are the most powerful resources on the earth, under the earth and above the earth. We have to empower them through value-based education and leadership"**. His leadership style is exemplary. He knows how to extract the best work culture from a team. He implemented a knowledge culture .Dr Vikram Sarabhai was his Guru. He use to tell to his ISRO colleagues that a leader should work with integrity and succeed with integrity. For India he had a vision 2030.

From a newspaper boy to people's president, remembering a legend, in his 3rd death anniversary of Dr A P J Abdul Kalam is relevant to the above subject and this article is a tribute to an able, noble and humble scientist from ISRO.

In his address at Asia net & ISRO meet in June 2008 at Rajiv Gandhi Bio technology Centre, Thiruvanathapuram answered questions from school children and he said **"When you wish upon a star, makes no difference what you are, anything your mind desires strongly will come to you. My message, especially to young people is to have courage to think differently, courage to invent, to travel the unexplored path, courage to discover the impossible and conquer the problems and succeed. These are great qualities that they must work towards. This is my message to the young people".** A P J Abdul kalam when he had worked with ISRO, he used to tell the young scientist here and students in the country to "look at the sky, we are not alone, the whole universe is friendly to us and conspires only to give the best to those who dream and work. Dreams are not those which come while we are sleeping, but dreams are those when you do not sleep before fulfilling them. To succeed in your mission, you must have a single minded devotion in your goal. If the country is to be corruption free and to become a nation of beautiful minds, I strongly feel there are three key societal members who can make a difference. They are the father, the mother, and the teacher. So my humble request to you all is be a role model to your children and perform your duties as an excellent father, mother & teacher". He wanted to be remembered as a teacher and his wish was to die while he was speaking to students. And, true enough, his wish was fulfilled. He collapsed and died while teaching the students about their role making the earth more livable. His last breath had in the class room of fifty students of IIM Shillonge in Meghalaya State . I too feel like many of you that RIP, in his case not "Rest In Peace". It should mean "**Return If Possible**". (RIP).

The man who ignited millions of young minds to dream more and dream big is no more with us physically. India is blessed with so many eminent personalities like him. Future generations of India, the 80 crore youth will be able to hold the head high in self confidence, and not turned down in self–doubt.

Transforming India's youth power in to enlightened citizens of India is generally named as Kalam's effect. That was his dream and India need to become a developed country. The bold decisions as President of India annoyed some political parties and hence he had not given a second chance as President of India. He, truly, was one of a simple personality. A rare gem from Rameswaram village became a Bharat Ratna and President of India, only by dedication to his assigned work, hard work and patriotism.

The top Ten from Dr A P J Abdul Kalam's vision list for the Country he loves for the vision 2030

1) A Nation where the rural and urban divide has reduced to a thin line.

2) A Nation where there is an equitable distribution of recourses, and access to, energy and quality water to all.

3) A Nation where agriculture, industry and service sector work together in symphony, absorbing technology, thereby resulting in sustained wealth generation leading to greater high value employment opportunity to all.

4) A Nation where education is not denied to any meritorious candidate because of societal or economic discrimination.

5) A Nation which is the best destination for then most talented scholars, scientists and investors from all over the world.

6) A Nation where the best healthcare is available to everyone, and communicable disease like AIDS/TB, water and water-borne diseases, cardiac diseases, cancer and diabetes are brought down.

7) A Nation where the governance uses the best of technologies to be responsive, transparent, fully connected in a high bandwidth e-governance grid, easily accessible and simple in rules, hereby corruption-free.

8) A Nation where poverty has been totally eradicated, illiteracy removed and crimes against women and children are absent and the society feels un alienated.

9) A Nation that is prosperous, healthy, secure, peaceful and happy and has a sustainable growth path.

10) A Nation that is one of the best places to live in on the earth and brings smiles to a billion-plus faces.

    According to Dr A P J Abdul Kalam thinking beyond our planet is an essential trait. The thought itself elevates the person. The person is transformed into a creative state. Creativity indeed is the foundation of discovery and inventions.

following space missions will give tremendous advancement globally. His Space vision 2030 for heavy lift rockets for inter planetary missions including-

- *Manned Space Mission to Moon & Mars and establishment of space industry.*
- *Cost effective space transportation systems using Hypersonic Reusable Vehicles, capable of placing up to 10Ttone class of spacecraft in GTO.*
- *Harnessing Space energy for power & drinking water.*
- *Developing Solar Sail for inter planetary missions.(Now solar aircraft Si 2 had completed its experimental flight successfully).*
- *Integrated Disaster Management – Role of space technology to be used.*
- *Refueling, Repair & Maintenance of Satellites in Geostationary Orbit in a cost effective way.*
- *Operational Navigational Satellites System (IRNSS).*
- *International "Youth Power" connectivity satellite.(Empowerment of youth using youth satellite).*

Dr Kalam, who, as the President, launched Vision Kerala 2020 before the Kerala Assembly in 2005. He had given 10 projects action plans for states development in the area including tourism, developing and protecting water ways, transportation using canals, deep sea fishing using satellites, Development and marketing of Ayurveda treatments and medicines, Starting Nursing courses to meet international demand, economic zone for NRI investors, value addition to rubber, coconut, cashew, space/defense requirements for industrial development etc .He use to visit the spiritual leaders of Kerala and talk about the role they have in his visions. His quote "Where there is right spirituality in the heart ,there is beauty in the character .Where there is beauty in the character there is harmony in the home, where there is harmony in the home there is order in the nation, where there is order in the nation there is peace in the world ".

Dr Kalam had a special place in his vision for space research. His quote "In 2020 the population rate in India will be 140 crore, in

2030 will be 150 crore, in 2050 population will be 170 crore. Space Technology has a role for extending the following to common man.

Energy, water, health care, education and employment potential are the important requirements for the people to live with prosperity. India required large needs technology inputs for meeting these needs. Space technology has a role. The vision of various space faring nations as well as discussion in various international forums by space experts suggest that space missions beyond earth are viable for sustaining the spirit of deep space exploration and for buildup of space infrastructure leading to space industrialization. Such missions would include bringing minerals and other special materials from Moon, Asteroids and Mars ,and also enable to moot the idea of a world space knowledge platform to launch a feasibility study for space solar power ,with partner nations providing seed funds for the four billion dollar project."

"The continuous clean energy provided by the space solar mission would lead to an era of peace, prosperity and abundance for all humanity. These missions would call for large mass flow into space and space industrialization and space exploration will expand using the new generation launch vehicles of 10 Tone class, The real value of space exploration for human advancement will occur only when mankind builds an optimal and robust design incorporating the latest technologies. We are still a long way from placing colonies on the Moon or Mars. Future missions promised to be even more captivating as a greater number of humans joined in the quest for space. For each of us Space research means something different".

Making Livable Planet-The Mother Earth is his last lecture at IIM, Shillong. Before completing the class and while performing as a teacher the true Karma Yogi Dr Kalam collapsed and died at the age of 84. A cardiac arrest ended his journey. A journey un matched in its width and heights. Either way he was one of the worthiest son of mother Earth. His way of living and vision may be the best message to youth in India for making India a developed country.

CHAPTER 1

KNOWLEDGE SHEET FOR BECOMING A SPACE SCIENTIST

Introduction

It is obvious that Dr, Kalam's parents never had a dream to perform him as a space scientist. It was his dream to come up from a poor family. To become a space scientist, you must have a dream for it. In the present generation, parents only have dreams and they work hard to full fill it through their children. Knowledge is power. Knowledge is a bunch of useful data used for creative purpose. Data can be useful or useless, that depends the user's need aspect.

Science & Technology are two different subject. Science deals with knowledge about theorems, hypothesis etc and can be learned from schools & colleges. Technology is the application of science and also called wisdom or "know how" obtaining only through experience. It is costly and time consuming. Most of the terms used in this write-up are technical semi- popular in style ,that is ,neither rigorously technical nor overly simplified for school level students.

The infinite space is full of countless mysteries which human being have been exploring in all centuries. Since the resources in Earth are depleting in a faster way it is necessary to find out alternate for energy, minerals, metals etc for exploration. To fly into space to understand better our home planet Earth from Space and utilize resources of space are the major challenges in the years to come. So Space research also inspires the younger generation all

over the World and this task required multi disciplinary and interdisciplinary global approach, together with nifty team work, to address the real world challenges for energy, drinking water, environment pollution, food, health and security of the nation.

Most of the people think of space as a nebulous region far above our heads with sun, moon, and stars and in day time blue sky extending out to infinity .But in the night dark sky.,(Star twinkle because of disturbances in our atmosphere). Space is actually black in color and a place where things happen , Rockets and satellite systems , Space Transportation System (STS), Human space flights, Moon and Mars mission etc. Even more than that those who spent their whole life for the inventions of rockets&, satellites for the whole Society and for the whole World **"Loka Samastha Sukino Bhavanthu"**. There are precious few Newton and Einstein among the readers of this Book. Most brilliance arises from ordinary people working together in extraordinary ways for inventing new products using advanced technologies; example mobile phone.

With the launch of the first artificial Satellite Sputnik a new technology called space technology was born. Actually it is not a new technology but a combination of different type of technologies. About hundred years after the death of Newton some imaginative authors had written science fiction stories about *Parakum Thalika* (Flying Saucer) and journey to moon and mars impressed by some of the fictional stories a few brilliance were trying to work out the real scientific basis of Space travel. Among these brilliance was a Russian School Teacher Konstantin Tsiolkovsky (1857-1935) who worked out the theoretical principles of Space travel using Rockets. His pioneering contribution in the field of Rocketry and Space travel made him the Father of Space Travel.

Today NASA-(National Aeronautics and Space Administration USA), ESA (European Space Agency ,France),ISRO,(Indian Space Research Organization) Arian Space Agency, French Guiana , JAXA (Japan Aerospace Exploration Agency &China are all the super powers in space. ISRO has put India on the world map of space powers. With an indigenous capability to design develop ,build, launch and operate its own launch vehicle and satellite even to Moon & Mars and a strong will power to apply this technology only to peaceful applications.

Vikram Sarabhai was the visionary behind the Indian space programmes. Now ISRO had proved its vast potential for application in Telecommunications, DTH, Television, Education, Meteorology and natural resources monitoring, Regional Navigation systems and interplanetary missions etc. I had been a part of this fine organization for more than Thirty three years ,blessed to work for the common man of India. To my experience, ISRO is having a good review system to every minute activity. This may be the reason for the success in space mission, a risky business.

Now we will see the legendaries who laid the foundation for the Space exploration

Konstantin Eduardovich Tsiolkovsly (1857-1935)

Konstantin Eduardovich Tsiolkovsly worked out many important equations relating important parameters such as specific impulse, thrust coefficient, area ratio etc. Many concepts, like regenerative cooling were worked out by him more than a century ago. He wrote the famous rocket equation. He said "Earth is a cradle of humanity but one cannot remain in the cradle for ever".

Dr. Vikram Ambalal Sarabhai (1919 – 1970)

Dr Vikram Ambalal Sarabhai (1919-1970) wanted India to pursue space science for the benefit of its people. Vikram Sarabhai became Atomic Energy Commission chairman in 1966 and later became chairman of Indian Space Research Organization. He initiated Indian launch vehicle programme in the early seventies. Sarabhai died in 1970 at the age of 52. He was a true pioneer and is rightly called the father of Indian Space Programme. He said "we must be second to none in the application of advanced technologies to the real problems of man and society" and " The development of a Nation is intimately linked with the understanding and application of Science and Technology by its people".

WHERE IS SPACE & WHAT ARE THE SCOPE FOR SPACE RESEARCH IN THE WORLD.?

Space is a place. Space begins somewhere above our heads, but how far? There is no universally accepted definition. Space scientist believe that Space starts at an altitude of 93km from earth. If you can go further at 130km stars can be seen and completely dark at all time with vacuum environment can be called as starting point of deep Space. ISRO is having deep space network station at Karnataka for the control ,guidance & simulation of inter planetary missions. Tracking of Mars Mission had done from this place.

Getting into Space is dangerous and expensive, but Space offer several compelling advantages for modern society. Space offers us abundant resources such as solar energy and extra terrestrial materials .Freefall environment enabling to develop advanced materials impossible to make in earth surface. To form certain new metal alloys, for example: we must blend two or more metals Aluminum& Steel in just the right proportion. Unfortunately, gravity tends to pull heavier metals to the bottom of their container, making a uniform mixture difficult to obtain in earth . But Space offers a solution. A manufacturing plant in orbit like International Space Station is used for doing experiments. This station literally falling towards the earth but never hitting the earth. This is a condition known as freefall. In freefall there are no contact forces on an object, so, we say it is weightless, making uniform mixtures of dissimilar materials possible at space.

The boundary of the solar system offers an untapped reserve of minerals and energy to sustain the spread of mankind beyond the cradle of earth. Space craft now use only one of these

abundant resource using solar panels. Uninterrupted limitless solar energy. But Scientists have speculated that we could use lunar resources, or even those from the asteroids, to fuel a growing space based economy. Lunar soil for example is known to be rich in oxygen and aluminum. We could use this oxygen in rocket engines and humans to breath.

Aluminum is an important metal for various industrial use. It is also possible that water ice may be trapped in craters at the lunar poles. In future Chandrayan -2 will do more experiment. These resources, coupled with human drive is used to explore the space .The sky is truly the limit.

Human pursuit of aerospace has learnt many lessons from nature and few adaptations have led to phenomenal progress. Insect flight uses complex unsteady aerodynamics. Dragonfly is a super maneuverable insect to study by project students in engineering college.

CHAPTER 2

ROCKET

These articles are targeted mainly to high school and engineering college level students for motivating in their basics studies to become a space scientist. Articles are given emphasis to 21 century future opportunities, challenges & perspective in global space industry for youth in India and abroad. The students must understand the basic subjects Physics, Chemistry & Mathematics since these three subjects' fundamentals and applications are complimenting each other in technology development and mathematical modeling. More examples to prove this statement, I will try to include in this article related to the basic theory and functions of rocket and satellite systems.

When students studied Physics, Chemistry & Mathematics in High school & Engineering College class rooms, they may have not been able to imagine their numerous applications in various fields. So the objective behind writing this article is to throw light on the possible application of the subjects they studied and illuminate their thinking process and to act as a motivator for young engineers in the country for future inventions like nuclear and electric rockets for inter planetary missions, wire les power from solar satellites, habitat in Moon & Mars, space tourism, reusable launch vehicle technology etc .

From my experience to work in the area of Space Technology, one has to pay attention even to the minute things. (Eg. The kite flies

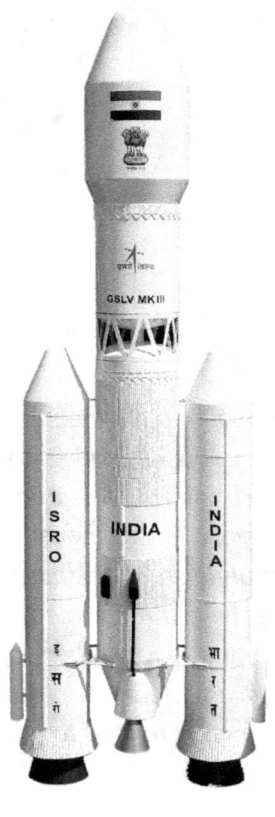

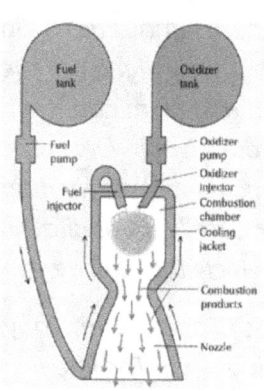

better because of its tail). Many people have seen apple falling down from the tree, but Sir Issac Newton, the father of modern science thought about it, and discovered the science behind it. As students are aware the Newton's Laws of Motion and law of conservation of momentum are the back bone of rocketry.

A rocket is basically a system that converts energy of combustion into force or thrust to move. The input mass consists of propellant and oxidizer. Rocket is a type of engine or prime mover that pushes itself forward by producing thrust. Unlike an aero plane engine, which operates within earth atmosphere by drawing in outside air to utilize the oxygen for fuel combustion and produce thrust. A rocket engine has to carry fuel & oxidizer to operate in space. This is the main difference between a rocket engine and aero plane engine. The figure below shows the cross section of a Liquid Propulsion Stage with Pump Fed Engine. In upper stages pressure fed engines and cryogenic engines are used. The figure below is a schematic of a typical liquid propulsion engine with fuel and oxidizer tanks. This is a pump fed engine with injector, combustion chamber and nozzle to develop required thrust. This engine is used in second stage of PSLV and four strap one motor of GSLV of ISRO.

Now we will study how the knowledge from school level physics, chemistry and mathematics applied in rocketry with examples: Rocket Equations (School Level).

A Launch vehicle needs a large velocity to get away from earth's pull to reach a specified orbit. This requires knowledge about, Newton laws of motion to calculate how high speed exhaust going in one direction pushes a vehicle in another. The simplest example of this is a toy balloon. What makes the balloon to go? Newton law says that for every action there is an equal but opposite reaction. When we let go of the stem of the balloon rocket propulsion causes the balloon to fly wildly in the room. When you blow into a balloon, you force air into it, making the rubber skin stretch, increasing the internal air pressure, and storing energy like a spring. When you let go of the stem, the air pressure has an escape route, so the skin releases, forcing the air out under pressure. Following Newton's law, as the air, which has mass, is forced out in one direction (the action), an equal force pushes the toy balloon in the opposite direction (the reaction).

Let's look at the action and reaction situation in bit more detail to see where the force comes from? Consider a person sitting on a trolley which moves along a rail with a load of stones. If he is initially at rest and begins to throw the stones in one direction, because of Newton's Law, an equal but opposite force will move him and the trolley in the opposite direction. Recoil of a gun is also an example, which we had studied in our school days.

To throw the stones, the person applies a force to the trolley. This force is identical in magnitude, but opposite in direction to the force applied the person and the trolley. However, remember the concept of conservation of linear momentum which you had studied in physics class. It says that the change in velocity of the stone will be greater than the change in the velocity of the trolley, since stone has les mass compared to trolley. The product of mass and velocity of trolley will be equal to the product of mass and velocity of the stones.

The stones leave at a rate which is equal to the mass flow rate, for a rocket it is mathematically represented by "m dot" and measured in units of kilograms per second. Linear momentum is always conserved, so as the momentum of the ejected mass of stones goes in one direction. The momentum of the trolley goes in the other direction. This basic principle produces rocket thrust. A rocket expends energy to eject mass out one end at high velocity in the nozzle, pushing the attached vehicle and satellite in the opposite direction.

Momentum change with respect to time is the force acting on a body and has the same units as force. The force acting on the rocket is defined as thrust. We also define a comprehensive term called effective exhaust velocity, C, that tells as how fast the high energy propellant is leaving the rocket. Newton's Law helps us to relate all above parameters mathematically as.

Equation-1 Fundamental rocket equation for calculating, Thrust.

F (Thrust of a rocket) = m dot C.

Where m dot=mass flow rate
C= effective exhaust velocity, as explained above.

Force is the product of mass and acceleration. Acceleration is rate of change of velocity with respect to time. This relationship should make sense from our trolley and stone example. The person can increase the force on the trolley, by either increasing the number of small stones he throws in a given time or he can throw big stones in a given time, i.e. higher mass flow rate m dot, or by throwing the stones faster, i.e. higher exhaust velocity C. Or the person can do both. This is a simple example to have a better understanding of a much complex working of a rocket system.

Of course, exhaust velocity for typical rockets are much, higher than anyone can achieve by throwing big/ small stones from a trolley example. For a rocket similar to the Space Launch System (SLS Americans new rocket for deep space explorations), and falcon 9 of space X, the exhaust velocity is about 3 km/s. As these high velocities are hard to visualize, it's useful to define power for unit time.

At lift- off, the Falcon 9 the two stage rocket of Space X company working for NASA with its 9 semi cryogenic Merlin engines produce 1.6 million pounds thrust (6672 k N in Space). That is approximately five 747 aero plane at full power during takeoff. Unlike aero plane a rocket power increases with altitude.

Thrust is an important concept of performance of a rocket. The thrust produced by a rocket depends only on the velocity of the propellant ejected (effective exhaust velocity) and how much mass is ejected in a given time (mass flow rate, m dot).To have better design calculations, equations are derived for the thrust produced by rockets. Towards this we need to introduce a new concept, the Specific impulse.

Specific Impulse will aid us to calculate the total velocity change of a rocket. Specific impulse value differs for solid, liquid, cryogenic, and semi cryogenic rockets. Configuration of a rocket and its efficiency mainly depend on Specific Impulse of the propulsion systems, provided in the rocket assembly.

SPECIFIC IMPLUSE - I sp.

Rocket produces thrust that pushes on a vehicle with rocket speed. Then what happens? If you push on a door, it opens. If you hit a ball with a cricket bat, it flies to the outfield. Returning to our person in the trolley example, note that to give the stones that velocity of a cricket ball, he has to apply a force to the stones like a hit using a bat.

Large force applied to an object over a short time produces an impulse. When the cricket bat hits the ball the impact seems to be instantaneous, hitting the ball for a fraction of a second produces the rate of change of velocity for a given period of time to the ball. Hence the acceleration to produce force is rate of change of velocity of the cricket ball example is also true in rocket propulsion.

To change momentum to attain a high speed, we can apply large force acting over a short time, like a cricket bat hitting a ball or smaller force acting over a longer time, like an ant moving a bread crumb in the garden. We define total impulse, as applying a large force on an object for some length of time. This result is the same as the objects change the momentum. Again, think about the bat hitting the ball. The muscles in our arms produce a force. We apply this force on a bat for a short time, which produces a total impulse on the ball, change its momentum, and drives it out over the fence.

Impulse works the same way for rockets as it does for cricket ball by an expert batsman. Our objective is to change the rocket's velocity and hence its momentum. Hence we must impart some impulse. This impulse comes from the thrust acting over a time interval. Although total impulse is useful for understanding the total effect of rockets thrust, does not give us much insight into the rocket efficiency. To compare the performance of different type of rockets we must know the value of specific impulse – one of the most useful terms in rocket science.

Specific impulse, I sp. will describe the cost, in terms of the propellant mass, needed to produce a given thrust on a rocket. In other words, specific impulse tells us "bang for the buck" for a given rocket. I sp. must be higher always, better in terms of the rocket's overall efficiency. The selection of propellants solid, liquid, cryogenic, air breathing and semi cryogenic etc depends upon the mission requirements and I sp the specific impulse.

Equation–2 for calculating the rocket efficiency the specific impulse, (unit is Seconds). $I_{sp} = \dfrac{F\ Thrust}{m\ dot\ g_o}$

Where I_{sp} = specific impulse (s) in seconds
F thrust = force or thrust (N).
m dot = propellant's mass flow rate (kg/s)
g_o = acceleration due to gravity on earth = 9.891 m/s2

    I sp represents rocket efficiency, the ratio of what we get (momentum change) to what we spend for (propellant). So the higher the specific impulse I sp the more efficient the rocket is.

    Earlier, we found the force of thrust in terms of the mass flow rate and the effective exhaust velocity. By substituting Equation – 1 into Equation – 2, we get another Eqation-3 Useful expression I sp.

Equation – 3- for calculating the Specific Impulse-I sp $I_{sp} = \dfrac{C}{g_o}$

Where

C = effective exhaust velocity (m/sec) g_o= acceleration due to gravity on earth =9.891 m/sec^2.

    Note g_o is a constant value representing the acceleration due to gravity at sea level, which we use to calibrate the equation. This, means no matter where ever we go in the universe, we humans will use the same value of g_o to measure rocket performance.

    As a measure of rocket performance, I sp is like the mileage rating given for automobiles. The higher the I sp, the more velocity change (ΔV), it will deliver for a given weight of propellant. Another way to think about I sp is that the faster a rocket can expel propellant, the more efficient it is. The velocity change is represented by (ΔV).

When we take a long trip in our car, we have to make sure that we have enough fuel in the tank to get there. This concern is even more important for a trip into space where no fuel pumps are there along the way. But how do we determine how much propellant we need for a given mission.

Naturally, some rockets are more efficient than others. For example, one rocket may need 100 kg of propellant to change velocity by 100 m/sec, while another may needs only 5o kg depending on the propulsion system. To figure out how much propellant we need for a given mission, we must have a relationship between the velocity change and the amount of propellant used. We call this relationship the ideal rocket equation. It tells us how much velocity change (ΔV) we get for a certain amount of propellant used and total mission cost.

The important Concept of the velocity change (ΔV) delivered by a rocket depends on its effective exhaust velocity© and the ratio of initial to final mass of the rocket. The higher the effective exhaust velocity, the more velocity change ΔV delivered for a given mass of propellant used. Mathematically now we will go for Equations-4 for change in velocity to put a satellite in the specified orbit and inclination, depending upon the satellite application. Escape velocity required for circular orbit is 8 km/sec, elliptical orbit 11 km/sec, for interplanetary missions 17 km/sec. This equation is very much useful for calculating the propellant required in satellite launching and future inter planetary missions and cost effective access to space industry for economical growth of the country.

Equation – 4   for calculating the velocity change. (m/sec)

$$\Delta V = C \frac{M_{initial}}{M_{final}}$$

Where ΔV = Velocitychange(m/s)
C = Effective exhaust velocity (m/s)
$M_{initial}$ = Vehicle's initial mass, before firing the rocket (kg)
M_{final} = Vehicle's final mass, after firing the rocket (kg)

    Equation – 4 is one of the most useful relationships of rocket propulsion. Armed with this equation, we can determine how much propellant we need to do almost anything, from stopping the spin of a spacecraft in orbit, restarting the engine for longer missions like Mars and to launch satellite to desired orbit and inclination, inter planetary mission for habitat construction in Moon or Mars etc. The difference between initial and final mass represents the amount of propellant used. ΔV is also a function of the effective exhaust velocity.

    This relationship should make importance because, as the propellant moves out of the nozzles faster there is an increase in the momentum. The rocket goes faster to attain the velocity to reach the specified orbit for its assigned functions. This will increase the life of the satellite and more revenue will be generated from it for several years in the orbit for its functions.

    We can substitute the definition of I_{sp} into the rocket Equation (4) $C = I_{sp} g_o$ derived from Equation (3), to compute the ΔV for a rocket, if we know the I_{sp} and the rocket's initial and final mass, velocity change can be calculated by another Equation-(5) as shown below. This will show you how mathematics is helping in technology developments designs.

Equation – 5 $$\Delta V = I_{sp} g_o \frac{M_{initial}}{M_{final}}$$

Where
ΔV = Velocity change (m/s)
I sp = Propellant's specific impulse(s)
g_o = acceleration due to gravity at sea level (9.81m/s2)
M initial = Vehicle's initial mass, before firing the rocket (kg)
M final = Vehicle's final mass, after firing the rocket (kg)

Flow chart below shows a simplified view of a rocket stage system.

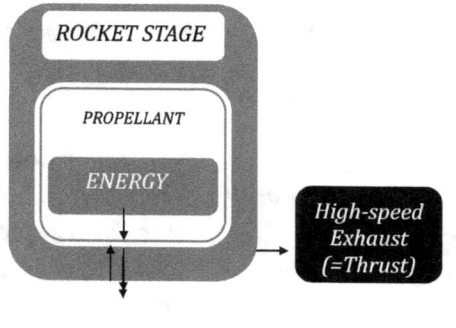

This Flow chart shows a simplified rocket system. Rocket Stage takes in propellant and energy to produce a high speed exhaust. Conservation of momentum between the exhaust and the rocket stage produces thrust.

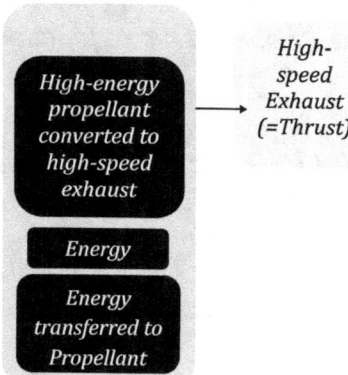

This Flow chart shows more detailed view of rocket-Stage. Energy first transfers to the incoming mass. This high-energy mass then converts to high-speed mass, producing thrust. This will give the required velocity change for the correct launching of satellites in to the orbit with cost effective.

Nozzle opening in the bottom of a rocket stage will create an unbalance in pressure at thrust chamber walls and with Newton's law of conservation of momentum between the exhaust and the rocket stage, the rocket will move in the opposite direction producing required thrust.

Rockets are the only vehicles powerful enough to reach space and carry people and equipments into space. Since the Soviet Union lifted the Sputnik-1 in 1957, and in 2018 the dangerous fundamental equations given in this article have remained the same for new inventions. Rockets have lifted thousands of artificial satellites into orbit around Earth. These satellites have taken pictures of Earth, monitored the weather, gathered information for scientific studies, and transmit communications around the world. Rockets also carry scientific instruments far into space to explore and study other planets. Since 1961, rockets have launched spacecraft carrying astronauts into orbit around Earth and to space conducting experiments and studies like International Space Station (ISS) of NASA as shown above. China progresses towards manned space station Tiangong 2 in to orbit 380 km above earth and will establish a laboratory in space in the year 2022. Shenzhou-2 manned spaceship will ferry two astronauts to this lab on 15 October 2016 for 30 days work in free fall environment. Russia started space tourism in a large scale. America started inter-planetary missions study using falcon heavy rockets in 2030.

In 1969, rockets carried astronauts to the first landing on the Moon using Apollo rockets. In 1981, rocket lifted the first space shuttle into Earth orbit. Space shuttle retired in 2011 and replaced by Falcon 9rocket of space X, made history in 2012, when it delivered Dragon in to correct orbit for rendezvous with the International Space Station (ISS). This Falcon rocket had failed several times resulting heavy loss to Space X private company at USA. NASA had published each success and failure with failure analysis in detail. Many more space achievements & failures had occurred in the area of space exploration in the world and in India ISRO is showing a better performance in a cost effective manner with more than eleven launches per year which includes GSLV Mk 3 heavy lift made in India rocket. Congratulations to my colleagues in ISRO.

Space is difficult and is a demanding environment. Space flight is an incredible challenge, but Scientists and Engineers learn from each success and each failure and designing heavy lift rockets like falcon 9. Science fiction movies like "space between us" going to be released soon by Hollywood will also give inspiration to youth for Man mission to Mars in 2030 by U.S.A.

The role of space is a strong enabler for social transformation, a catalyst for economic development, a tool for enhancing human resource and to strengthen national security. Even though the technological improvements and scientific temperament in space technology had increased, the human errors towards the dangerous fundamentals cannot be neglected. This industry is handling large quantity of highly explosives chemical propellants. In the case of disasters due to human error and machine failure the shock and impact dynamics of these high explosives will end up with direct loss of assets like facilities and human resources. Also have to face indirect loss like reputation, schedule slip, morality and credibility of the organization. Hence all the space flaring nations will give extra care for quality, reliability and working standards with safety measures and no compromise is allowed in fallowing safety standards.

America, China,& Japan are progressing in space application in the world very fast. In India, ISRO started to launch one rocket per month and each launch more than one satellite. In some launches ten to twenty five foreign satellites together to reduce the launch cost. Chandrayan 2 & heavy lift rocket GSLV Mk 3 are future goal of ISRO. For communication ISRO is having eleven satellites operational with 236 transponders as on December 2016. Now the transponder requirement in India is 400. This shows the present demand itself is high and it will goes on increasing with

Infinite Space & Unlimited Excitements

GSLV Mk 3

population and applications. In remote sensing area three satellite only operational and requirement is ten in sun synchronous orbit. All demands are people centric hence ensuring safety, security and quality is very important in space based systems.

If all the satellites of ISRO are declaring strike we cannot imagine the problems in India in banking, techno park, navigation systems in air ports, telecommunications net work, defense, direct to home DTH TV disaster warning ,mobile phone , internet on line application. ATM etc . Throughout this book I will focus primarily on the practical aspect of space - what it likes, how do we get there and explore and how do we use space for the development of India? Challenges & perspectives we will discus in Mission to Moon & Mars.

THE SPACE IMPERATIVE

Going to space is dangerous, risky and very costly. But space offers several compelling advantages for the development of a nation. It offers a global perspective- the ultimate high ground, a clear view of the heaven, a free fall environment for developing advanced materials and alloys impossible to make on earth, abundant resources like solar energy and extra terrestrial materials and a unique challenge as final frontier. The alarming thing in earth is the pollution and the climate change. Monitoring the earth from space globally is mandatory to address the global warming and climate change issues in the world. Wild life and deforestation is a global issue. Human beings are the only homo sapience spoiling the equilibrium of the environment. We are destroying our soils, rivers and oceans by putting waste and plastic, burning waste and plastic destroying breathing air. Soil is not charged with rain water and water is allowed to flow to oceans. All reveres are polluted with waste. Accumulation of waste in the environment and no water in the soil destroys trees. Trees are the lungs of our land. Trees are purifying the air and giving fresh strength to Human beings and habitates

Animals' kills for their need but humans' kills for their greed. Kerala must note that all this greeneries , animals and rivers of Kerala may vanish in a matter of time. If we don't start acting now. Half the world will turn to desert soon and rain will go away like Vyanad in Kerala . Some area may become flooded. Global warming and Climate change are byproduct of human greed. It can produce cyclonic wind with unpredictable wind velocity in coastal area of India. Kerala is also not free from these phenomena. The solution is very complex due to the increasing population in India especially in Kerala. . The people hesitate to live in harmony with mother earth and her habitats   and keep earth clean and ever green.

Various Types of Rocket Engines used world wide

This chemical rockets can operate in outer space, where there is almost no air very low temperature and deep vacuum. A rocket can produce more power for its size than any other kind of engine. For example, the main nine rocket semi cryogenic marline engine of the **Falcon 9** manufactured by a first commercial

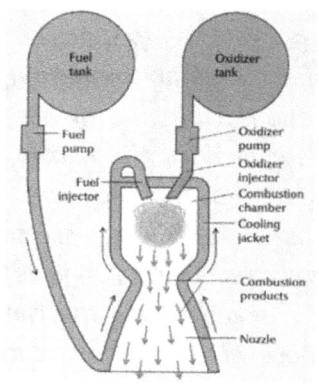

Infinite Space & Unlimited Excitements

launch vehicle company called **Space X** now working for NASA weighs only a fraction as much as a railway engine, but it produces approximately thrust of 40 railway engines.

Falcon 9 main stage booster is reusable for 10 to 20 launches, possibly up to 100 with minor refurbishment, made history on 08-04-2016 by breakthrough landing in a floating platform in sea. Falcon 9 successfully docked Dragon capsule to International Space Station (ISS) of supplies of weight 3175 kg. Company Space X is concentrated to make a perfect engine named as merline engine and clustering these engines to make powerful rocket. They are manufacturing this engine in a large scale instead of making several type engines for a launch vehicle to reduce cost and increase system reliability. Space X the American privet company had realized 9x3=27 clustered engine for a pay load capacity 64 tones to low earth orbit and 17 tones to Mars named as Falcon Heavy rocket. Falcon Heveay is cost effective and proved as a work horse of NASA even for Mars missions. . .

Falcon 9

Launch vehicle used for injecting satellite in to space is a specially designed multi staged rocket system. It provides the necessary velocity increment to get a satellite in to space. The inertial average velocity is 10 km/sec. At lift-off the rocket blasts off to gain altitude rapidly and get out of the dense atmosphere. Which slows it down due to drag by the influence of atmospheric air . When it gets high enough, it slowly pitches over to gain horizontal velocity, and this horizontal/ tangential velocity keeps a satellite into orbit for various applications. Orbit details for various applications I will explain in chapter 3 page 32 onwards (Satellite and orbits).

It is difficult to do the specified orbit raising activity with a single stage rocket due to technology limitations. Launch vehicle consists of different stages like solid stage, liquid stage, semi cryogenic & cryogenic stages etc .Each stage based on the type of fuel used imparts incremental thrust as they burn out in succession. It is like runners in a relay race, passing the batons handing over to another. Normally two, three or four stage rockets are configured to form a launch vehicle.

Rockets can be designed as long-range missiles. Thus it aids in the country's deterrence capabilities. Long range missiles can have a strike range of over 5000 km and can carry a nuclear warhead of over 1000 kg .When a rocket fired horizontally it is called a missile, used by the defense for the country strategic strike capability. When they are fired vertically it is called progress of the country in the area of Science & Technology applications. Powerful Launch vehicles are used to lift, artificial satellites, humans, laboratories, rovers, robots, powerful telescopes into outer space and operation of shuttle service between earth and international space station (ISS).

Rocket engines generate thrust by expelling gas which is produced during combustion. Rockets produce thrust by burning a mixture of fuel and an oxidizer, a substance that enables the fuel to burn without drawing in outside air. Such rockets are called chemical rocket because burning fuel is a chemical reaction. The fuels together with oxidizer are called propellants. (Earth storable & cryogenic propellants which required chill down temperature storage hangers and double walled containers to transport).

A chemical rocket can produce great power, but it consumes propellants rapidly. As a result, it needs huge amounts of propellants to work for even a short time. (Normally, the reaction is hypergolic). Chemical rocket engines get heated as the propellants combust. The temperature in some engines reaches 3300 degrees Centigrade, much higher than the temperature at which steel melts. The working pressure also reaches up to three digits. This makes the systems more complex to handle. Awareness of special materials, propellants and experiencing it is mandatory for personal safety & facility management.

Aero plane engines also burn fuel to generate thrust. Unlike rocket engines, however, jet engines work by drawing oxygen from the surrounding air is normally called as air breathing engines. For hypersonic launch vehicles, the use of an air-breathing propulsion system with supersonic combustion is a key technology that can be used in the atmosphere to go up to an altitude of 40 KM. To reduce cost this type of technology can be introduced in future rocket design.

NUCLEAR ROCKET PROPULSION

Scientist & Engineers have also developed rockets that do not burn chemical propellants. Nuclear rockets use the heat energy of a nuclear reactor, a device that releases energy by splitting atoms. Some proposed designs would use hydrogen as propellant. The rocket would store the hydrogen as a liquid. Heat from the reactor would boil the liquid, creating hydrogen gas. The gas would expand rapidly and push out from the nozzle producing thrust as shown in the figure below.

Future Rockets will be semi cryogenic liquid engines using liquid oxygen and purified Kerosene (House wife propellant). Space X Company in USA are now designed the biggest Liquid Rocket in the world having 27 Merline Engine clustered together Semi Cryogenic. India is also under development stage of heavy lift Rockets.

SIMPLE DIAGRAM OF A NUCLEAR ROCKET

The exhaust speed of a nuclear rocket might reach four times that of a chemical rocket. By expelling a large quantity of hydrogen, a nuclear rocket could therefore achieve high thrust. However, a nuclear rocket would require heavy shielding because a nuclear reactor uses

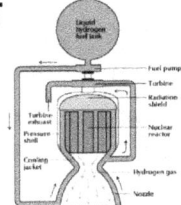

Image credit:
World Book diagram by
Precision Graphics

radioactive materials. The shielding would weigh so much that the rocket could not be practically used to boost a launch vehicle. More practical applications would use small nuclear engines with low, continuous thrust in order to reduce flight times to Mars, Moon or other inter planetary missions. This low thrust engine can also be used for the North South station keeping, NSSK operations of a satellite in orbit. A nuclear rocket uses the heat from a nuclear reactor to change a liquid fuel into a gas. Most of the fuel flows through the reactor. Some of the fuel, heated by the nozzle of the rocket, flows through the turbine. The turbine drives the fuel pump. Nuclear rocket developers must also overcome public fears that accidents involving such devices could release harmful radioactive materials. Designing nuclear rockets, the challenges in safety management also to be addressed, Engineers must convince the public that such devices are environment friendly.

ELECTRIC ROCKET PROPULSION.

Electric rockets use electric energy to expel ions (electrically charged particles) from the nozzle. Solar panels or a nuclear reactor can provide the energy. A simple diagram is shown here.

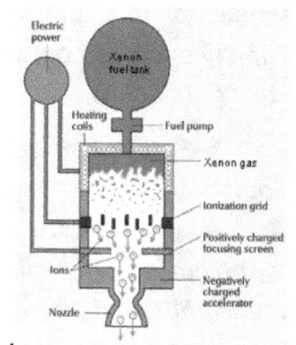

In one design as show below, xenon gas passes through an electrified metal grid. The grid strips electrons from the xenon atoms, turning them into positively charged ions. A positively charged screen repels the ions, focusing them into a beam. The beam then enters a negatively charged device called an accelerator. The accelerator speeds up the ions and shoots them out through a nozzle as shown above.

An ion rocket is a kind of electric rocket. Heating coils in the rocket change state of a fuel, such as xenon, into a vapor. A hot platinum or tungsten ionization grid changes the flowing vapor into a stream of electrically charged particles called ions to eject from the nozzle to produce thrust.

The exhaust from such electric rockets travels extremely fast. However, the stream of xenon ions has a relatively low mass. As a result, an electric rocket cannot produce enough thrust to overcome Earth's gravity. Hence it cannot be used as a booster rocket in first stage. Electric rockets used in space must therefore be launched by chemical rockets as booster in first stage. Once in space, however, the low rate of mass flow becomes an advantage. It enables an electric rocket to operate for a long time without running out of propellant in interplanetary missions like Mars.

The xenon rocket that powered the U.S. space probe Deep Space 1, launched in 1998, fired for a total of over 670 days using only 160 pounds (72 kilograms) of propellant. In addition, small electric rockets using xenon propellant have provided the thrust to keep communications satellites in position in the specified orbit for years.

Another type of electric rocket uses electromagnets, rather than charged screens to accelerate xenon ions. This type of rocket powers the SMART-1 lunar probe, launched by the European Space Agency. India is also developing electric rocket for Mars missions & satellite attitude and orbit control systems AOCS Engines, where more precision required. More R&D is required to reduce the cost and system reliability towards long life of sattelites.

The miniaturization of components and robotics will get extra scope in space technology R&D. ISRO semi-cryogenics R&D is in progress.(Cryogenics is the study of low temperature physics dealing at temperature below minus one fifty degree centigrade). In semi-cryogenics rockets liquid oxygen and purified kerosine are used as propellents.

SATELLITE AND ORBITS

Newton had predicted that if a body were thrown from a top of a high mountain with a speed of 8km/sec in a direction parallel to earth surface, it would become a satellite of the earth. This injecting speed is called escape velocity for circular orbit (Ref: Page No. 20).

Necessity is the mother of inventions. When value added Service is obtained from Inventions and Discoveries, it is called as an Innovation.

Satellite is an innovation. Value added service is now getting from satellites and launch vehicles. The present rate of placing one gram to space is approximately 20 US Dollars. Most of the innovations in defence & communications areas, are the spin of Space Technology developed by NASA, ESA, China, Japan, Russia, India etc. Value added service and cost effectiveness from Missiles & Mobile Phones services are examples of spin of space technology.

Scientific temperament started since Soviet Union had launched the sputnik satellite in 1957.Cold war between America & Russia acted as a catalyst for the advancement of space technology. Earlier the discovery of Electricity had helped industrialisation. Now without electricity people cannot live. The same conditions are going to come in the case of Space technology in the area of satellite applications. Chances of cold war between America and Russia again can't be ruled out in the present relation of U.S.A and Russia.

ISRO had obtained the Gandhi Peace award for its contributions in many areas to bring about socio-economic transformation in the country. Set to act as a force –multiplier in implementing more than 170 projects which include defence projects to qualify India a member of Missile Technology Control Regime (MTCR), an international agency.

Air ports authority for GPS-Aided and Geo-Augmented Navigation (GAGAN), INSAT,GSAT, Tele Communications Satellite spectrum, Remote sensing , Indian Regional Navigation Satellite System (IRNSS),Weather forecasting, DTH TV, Mobile phone , Internet, core Banking & ATM satellite connectivity , Railways to get on line G P S satellite imagery for improving safety & efficiency ,Wi-Fi in home, office ,.Trains etc, where Satellite technology can be used for improving the living conditions of common man.

The common man will appreciate, when they see the news paper and TV that, ISRO had frequent successful launches of PSLV, GSLV from Shriharikota. Few of them may think that ISRO is showing their power in the technological advancement and getting appreciation from the public & the Government.

ISRO missions are "Not for appreciation". Each launch is for a different end use as explained above and the satellite injected in to the orbit are for different applications. Some of the launching is commercial launches for foreign Countries satellites like USA, Singapore, etc. This will help to develop industrialisation to increase the GDP of the country. Indigenous development and production of Rockets & Satellites components with Indian industry made ISRO missions Cost effective.(eg. Interplanetary Mars Orbited Mission, Moon Mission etc.). India can establish space industry using youth power, with international co-operation.

Like NASA most of the "know How" developed by R&D organisations has to be transferred to Indian Industry for mass production for faster, better & cheaper with Zero Defect with Zero Effect to environment. Objective also be **"MAKE IN INDIA"** programme.

Earth rotates at a fixed rate at fifteen degree per hour (360^0 for 24 hrs). We can use this rotation as a clock to explain the orbit period. The earth spins on its axis at 1600 K M per Hour at the equator. The satellites are orbiting above equator, and moon is the natural satellite orbiting the earth.

SATELLITES ARE ORBITING THE EARTH (WITH DESIRED ORBIT AND INCLINATION)

Rockets inject satellite in to the desired orbit and inclination as shown in the figure. Orbit details we will be discussing later. As we discussed, each mission is having definite orbit and inclinations. Then only the mission is called as a text book mission. Space is hard, but worth it, hence countries take risk to push the boundaries of human achievements. Satellite launching is incredibly difficult; any error will seriously affect the performance and life of the satellite, normally the average life is 12 to 15 years. If the rocket fails to put the satellite in orbit the mission fails fully.

Hence rocket and satellite manufacturing and production required efficient configuration and Design, powerful propulsion systems, Structure Design and Engineering, Correct Specification Definition, Load Analysis, Environmental aspects like vibration, acoustics and thermal load Analysis, Factor of safety, Assembly & Integration, Testing and Analysis, Material choice, Fabrication methods, Manufacturing and production constraints, out sourcing to industries for space industry development in India etc.

Launching of satellite required Navigation, Control and Guidance System, vehicle health monitoring system, Telemetry, Tracking and Command System, Power Modules System, high altitude Simulations, Robust Control Electronics, Digital Autopilot system, Inertial Navigation System, Lithium Iron Battery and plant with solar pannels, Composite pressure vessels for reducing the weight and to carry propellants, Launch pad, propellant production and storage, failure analysis, measurement systems, trouble shooting mechanisms, project management, programme planning and evaluation, HRD, Man power, Machine, Money, Management tools for quality, reliability, Leadership, Team building and co ordinations of activities in various centers, cost effectiveness market survey for global tenders etc.

WHAT ARE TRAJECTORIES AND ORBITS?

A trajectory is a path a rocket / satellite follows through space. When it reaches the space Satellites resides in an ORBIT. Orbit is a fixed racetrack, on which it travels around earth. Like planets Orbit the Sun, we can place satellite in to orbit around earth, Similar to a car race track, orbit usually has an oval shape depending on the mission the orbit size, shape, and orientation will vary. Orbital mechanism is a vast subject, now we are discussing only fundamentals and details of orbits used for various applications like communication, GPS remote sensing defense, navigation etc. creating value from space.

ORBITS

(A) Depending upon the space applications, Orbit can be devided into three categories

(1) Equatorial orbit- in which the orbital plane lies in the earth's equatorial plane (36,000km away from earth).

(2) Polar orbits- in which the satellite move from pole to pole of the earth (90 degree with respect to earth equatorial plane).

(3) Inclined orbits – in which the orbital plane is at an angle to the earth equatorial plane (angle depending upon the mission objective and purpose of satellite).

(B) The altitude by which the satellite moves in space for its desired functions, mission objective and purpose can be divided into seven categories.

(1) Low earth orbit (LEO)(250km to 1000km) used for imaging, astronomy research.

(2) Medium altitude orbit (MAO) (100km – 20000km) used for earth imaging, navigation, astronomy and communication etc.

(3) High altitude orbit. (HAO)- (20000km and above) functions same as MAO.

(4) Geo- synchronous orbits (GSO). The orbit with the same speed of Earth rotation rate of 24 hrs with functions same as HAO.(20,000 Km and above).

(5) Geo stationary orbit is also a geo synchronous orbit(GSO) with an inclination of zero degree in the equatorial plane .(The satellite looks stationary over the equator) . when viewing from earth surface satellite is acting like a mirror. Whatever signals sent from earth is reflected to various locations in earth . This is useful for covering both the hemisphere of earth. Communication satellite and metrological satellite are Geo synchronous type. Injection of

Infinite Space & Unlimited Excitements

the satellite in to Geo stationary orbit is difficult and it requires lot of energy from the rocket and hence not economical. Hence rocket will inject the satellite into an intermediate Transfer orbit GTO, from GTO Satellite propulsion will take to GSO step by step using LAM (Liquid Apoge Motor fitted in satellite).

(6) GTO (Geo Transfer Orbit) Its nearest distencence to earth is called perigee in the range 180km to 650km. The farthest distance to earth is called Apogee is 36,000km. Inclination will be 7 degree from USA Kourou (NASA) and 19 degree from India. Shriharikota (ISRO) .

(7) Super Synchronous Transfer Orbit (SSTO) 36,000 km above.

In the case of satellite moving around earth, follows the same rules as the motion of planets around Sun in our Solar system, a famous scientist Johannes Kepler (1571-1630 AD) observed the motion of the planet and formulated three laws .

(1) The planets described ellipses with the sun as focus.

(2) The line joining the sun and the planet sweeps out equal areas in equal time.

(3) The square of period of revolution of the planets is proportional to the cube of their mean distance.

The concept of GSO was found in 1945 by the famous scientist Sir Arthur Clarke. He discovered that the period of a satellite at 36000 Km attitude will be equal to the time period required for earth to complete one full rotation. These two Scientists discovery helped to develop satellite technology in the area of communication, meteorology ,remote sensing, interplanetary missions etc.

The geostationary orbit GSO is defined as a circular orbit at an altitude of 35,786km above earth and in the orbital plane of earth equator (inclination angle ,i=0). A satellite in such an orbit will have an orbital period 23hrs 56 minutes 04 second. This is equal to one sidereal day of earth. The satellite in Geosynchronous orbit will move in exact synchronous with earth and appears stationary to the point vertically below it on the earth..Relative angular velocity is zero with respect to the observer to have continuous communication.

GSLV is used to put the satellite in GTO say 200 KM by 36000 KM with an inclination of X degree. (That is why GSLV is called as Geo Synchronous satellite launch vehicle). From GTO Satellite propulsion system LAM-Liquid Apogee Motor fitted with satellite is used for orbit raising to 36000KM by 36000 KM in GSO with an inclination nearly Zero degree, GSAT will be placed above the equator at a particular longitude North South station keeping is carried out by AOCS-Attitude and Orbit Control System using 22 N, 10 N thrusters fitted with the satellite.

Sun Synchronous Orbit SSO is defined as the orbit in which the orbital plane rotates in a year in unison with the one revolution per year apparent motion of the Sun as explained below.

The advantage of SSO is that earth observation condition can be kept with a constant solar incident angle. There is 365 days in a year. That means the Sun goes over a given spot at Thiruvananthapuram 365 times a year. If you draw a circle around Sun and divided in to 365 spots, each spot is "One Day". (360 degree of Earth revolution or we are rotating 15 degree/Hour).SSO Satellites orbit the earth in North-South direction of earth so as to take photograph of earth, when it is illuminated maximum by the Sun at particular region say Niagra Falls. The orbit time is 100 minutes. The satellite take picture at 10.30 AM. When region will be illuminated maximum by the sun light.

Satellite fall under the following categories such as Communication satellites-, TV/Radio Broadcasting, Meteorology, Navigation, mobile

communication, Remote sensing satellites-low earth polar orbit for various services, Science satellites for planetary, interplanetary exploration and Technology satellites for various other purposes. The market for satellites consistently grow during '90s with the planned but short-lived new concepts such as low earth orbit COMSAT.

The principal COMSAT Builders are the Hughes space systems ,Boeing, Lockheed-Martin and Space Systems/Loral in US and Astrium, Alcatel, Alenia Spazio in Europe. Today, the COMSATs provide life as high as 14-15 years. Recently, Intersputnik the Russian organisation has planned for 500-1000 kg COMSATs to place between 5-12 transponders with 5-7 years of life providing cost effective launch/ insurance/ sat development/ adaptability to changing demand/exclusivity for a particular market etc.

Similarly, the remote sensing satellites in the low earth polar orbit forms the second segment of satellites generating a huge commercial market in terms of imageries and value added services to global users. The technology has improved from 1 km resolution to presently in meter/sub-meter resolution with TES, IKONOS and others. They find numerous applications such as ice surveillance, crop monitoring, disaster assessment, forest applications including mapping and geological exploration, and coastal zone, ocean monitoring, urban planning, water resources etc. The Canadian Radarsat, French SPOT, Indian IRS series, European ERS and the US Landsat are the notable ones in this category. ISRO and Keltrone had signed an agreement for mass production of Navigation Satellites-NAVIC-DAT. Distress alert transmitter for the fishermen which detect distress beacons for search and rescue in the sea in disaster situations.

THE GLOBAL POSITIONING SYSTEM (G P S)

*S*atellite have been used for navigation-determining where you are and where you are going. Also offers incredible military and civilian applications like crime detection by police. For this we need to know, where the satellite need to point its instruments and antennas as shown for latitude, logitude, altitude and time.

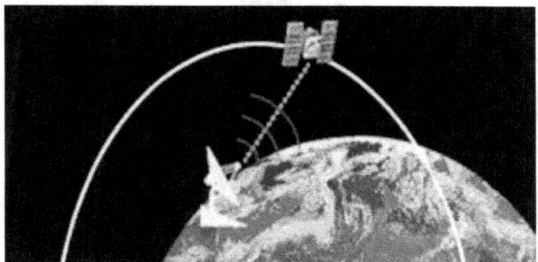

GPS requires a constellation of satellites to do the job. What are the jobs GPS will do? Navigation work in aerial, marine and terrestrial. Disaster management, Vehicle tracking and fleet management, Integration with mobile phone, Precise timing, Mapping and geodetic data capture, Terrestrial navigation aid for pilots and travelers, Visual and voice navigation for fisher men and drivers.

For doing the above job satellites have 24 hours duty without any holidays and leave. Every point on earth is viewed by at least four GPS satellites at a time. GPS is the trade name of a constellation consists of 24 operational satellites introduced by USA. All these Satellites placed in six symmetrical Orbital plains at

20,000 km above Earth in 12 hour orbits inclined to 55 degree to equator. Other GPS Systems are Glonass of Russia, Galileo of Europe is yet to be operational, Beidon of China for regional coverage and finally IRNSS of India-Generic name is National Navigation System of the country (NAVIC).

Indian Regional Navigation Satellite System (IRNSS) consisting of 7 satellites in orbit and 2 on the ground as standby for regional coverage area extending up to 1500 KM from India boundary. IRNSS is at par with American GPS in accuracy 20 Meters of position information service for Standard Positioning System to public (SPS)& Restricted Service (RS) to Defense Air ports,Railways, Police etc. Access to foreign government controlled GPS can be restricted and disturbed in hostile situations like war. ISRO also proved as a pioneer in Rocketry and Satellite manufacturing and launching in a cost effective way participating make in India policy to increase GDP of the Nation. Antrix Corporation is the commercial part of ISRO at Banglore.

ENGINEERING ASPECTS AND OTHER USES OF GPS

In order to fix any position on the Earth a specially made GPS receiver connected to a computer can be used to track the GPS satellites and calculate the position in few minutes this facility is very much helpful to human activities like land survey, water resources etc. Building area calculation in metro cities, Navigational reference for AIR / SHIP transportation industry, Adventures in space tourism, Ocean survey, deep sea fishing, traveling by Aero plane, ship this GPS technology is very helpful for accurate navigation. Equipment landing and takeoff for Aero plane in busy air ports and adverse weather conditions, Defense purpose GPS technology has been found to be very useful for coast guards, fighter plane, war ship etc. GPS system will help for targeting missiles for its better utilization during war for accurate position, velocity and time thus serve as eyes and ears in the sky.

HOW TO USE GPS?

Now in developed countries several cars fitted with GPS receiver, this receiver tracks the satellites and determine the location for the vehicle in terms of latitude and longitude . This data will map the location in the memory of the computer inside the car and calculate a route to the destination entered in the computer. As one drives to the selected route the system continuously provides guidance by Showing route on the map and displayed on the screen provided inside the car the same car is fitted with a gyroscopic sensor and a speed sensor keep track of the direction and speed of the vehicle all the time. And this system jointly drives the car forward. The user of the car has to enter the destination and required speed to the computer before starting the journey for getting perfect timing.

With the increasing of vehicle and heavy traffic in City roads in future GPS controlled vehicles will be the solution . The figure below shows how the satellite is getting perfect timing from highly accurate rubidium atomic clock fitted in the satellite for navigation service signals to users and synchronizing the satellite and the receiver fitted in the car, aeroplane, ship etc.

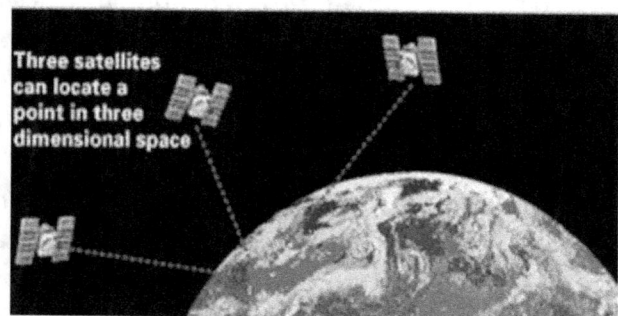

Navigation system provides visual and audio instructions to the pilot and the pilot need not take strain by looking forward during his journey. This GPS also provides location details in the ground for automatic parking, details of petrol pumps, hotel, rest house, hospitals etc on the way. The party can order the items one or two hour before during the journey and the item will be ready when they reach the hotel provided the location program is done correctly.

GPS Satellites are used as eyes and years of military.These satellite can obtain radio/telephone talk of enemies and can take photos from space most of the American satellites are at LEO for a shorter period. This satellites are called key hole satellites (KH). Using a high resolution camera and radar military operation can done precisely day and night and in cloudy condition. Nuclear test in space is not allowed so far. Satellite will detect the radiation from missile and give early warning for counter attack. Star wars Programmes are strongly criticized. Deploying weapons and laser jet in space is against of outer space treaty sighed on 1967. Space can only used for peaceful purpose in the world with the existing rules.

Earlier, only land phones are available, now mobile phones are used anywhere , anytime with our own phone numbers to call anybody on earth. A constellation of 66 nos. Satellites are called Iridium System in LEO are used for mobile telephones global coverage.

Satellites can track cyclonic storms continuously and help in predicting the probable locations of their land fall well in advance. Thus they are very useful for early cyclone warning. The picture shows a tropical cyclone formation in Bay of Bengal.

THE INTERNATIONAL SPACE STATION (ISS)

Infinite space and unlimited excitement also can be man made. The inspiring innovations of Space Technology can do wonders with international collaboration. The International Space Station (ISS) is an inspiring innovation of Space Technology. The largest international scientific and technological endeavour ever undertaken, is taking final shape in space. USA, Canada, Japan, nine members of the European Space Agency, and Russia are international collaborators. It is a permanent laboratory to be used where gravity, temperature and pressure can be manipulated in a variety of scientific and engineering pursuits in ways that are impossible in ground-based laboratories. A laboratory to make new advanced industrial materials, ball bearings, communications technology, and medical research and production of life saving nano medicines, crystals etc. are the major use of ISS, a working space laboratory.

The below picture shows the international space station consisting of seven research laboratories and manufacturing units in low earth orbit (LEO). This laboratory is orbiting on the top of our head in space. It is an orbiting station 338 km above Earth located in the south pacific will be the largest space station so far. The ISS will be seven storied with a total weight of about 500 tones and will be as large as two football stadiums. Both end huge solar panels are fitted for power generation to run the laboratory, 110 k w from the Sun.

Infinite Space & Unlimited Excitements

The below picture shows photos of ten cosmonauts working in the ISS laboratories as a team from different countries. Total volume of one laboratory will be equal to one747 jumbo jet aero plane. This size seven laboratories are now operational in ISS. One lady sitting in front row is Mrs. Sunitha Williams from India.

As you are aware Miss Kalpana chowla from India had died in February 2003 in a rocket blast accident of Colombia while returning to NASA from International space station. After the explosion in the re entry Space Shuttle Colombia causing the death of Miss Kalpana Chowla, NASA cancelled their flight and only the Russian rockets Soyuz and proton had used for the construction of ISS. The Space Shuttle flights were resumed in 26th July 2005 and on June 2nd 2008 delivered a bus sized Japanese laboratory named –KIBO Pressurized module to expand research and replaced badly needed parts for troublesome toilets used by the cosmonauts. Space shuttle retired in 2011 and falcon rocket made by a private company named space X in USA is started shuttling from 2015. Flacon carries instruments and food items, medicine etc for the cosmonauts. Space x falcon rocket

cosmonauts working in the ISS laboratories

failed two times to the mission to ISS. Recently in 2016 dragon space craft docked with the orbiting station and U S robotic specialist told ISS crew members that "you have a beautiful house in space to live and conduct experiments in zero gravity" the US crew members are staying more than three months in this ISS. ISS has been always occupied by a few cosmonauts who are periodically exchanged with those visiting from Earth. From Florida NASA had operated several flight to ISS. Space Tourism also possible.

The ISS is now ready for doing experiments in zero gravity or free fall conditions. To reduce cost ISS is working with international cooperation. India is yet to be a part of these missions. A unique advantage in an earth orbiting space station is the situation of weightlessness or micro gravity environment. In all earth-based experiments, gravity plays a major role. NASA conducting various experiments in the lab in several days as a team from different countries including Mrs Sunitha Williams from India.

Some costly life saving drugs like Interferon which is used for the treatment for HIV, hepatitis, Cancer etc is developed in ISS. Life saving drugs factory can be established at ISS since combinations of minute molecule mixing is easy in zero gravity environments. Nano medicine monopoly of USA is still exist for producing and fixing the cost of medicines internationally. Solar power is available in plenty in space there is no question of power failure for establishing manufacturing high quality metal ball bearings which reduce friction and noise levels to the vanishing point. These ball bearings can be used for jet engines of aero plane, helicopters, large radar antennas and large cargo aircraft wing support, propeller of big ships etc. In the absence of gravity, alloys of two materials having different densities like steel and glass can be made for optimum strength to weight ratio for manufacturing structures of passenger aero planes and satellites. Space tourism from earth and stay at ISS also is possible in future. Now some people had registered their name with NASA for trip to Moon &Mars. The challenges I will explain in Mission to Moon & Mars page no. 68.

Expansion of the ISS will continue with the delivery of big laboratory such as Harmony, KIBOW, Columbus etc. it is a joint venture. Harmony made by as a team Italy and KIBOW by Japan and Columbus by European Space agency. Future programs of augmentation of laboratories are planned. Total seven laboratories by sixteen countries international cooperation. Unfortunately India is not a member of this team so far.

In India, we only read about in the media about corruption, banking scam, death, sickness, terrorism, crime etc. Other countries in the word are also not free from these evils, but the media will never give more propaganda adverse to the nation and did not allow king fisher Vijay Mallayyia type people to grab huge public money and escape from the country safely. Unless India stands up the world and control the above evils , no one will respect us and make partnership with us. We are very good in application of softwares. But we are not good as developers. We have to improve a lot in software development In India.

In this scientific world, fear for implementing new has no place to live. Only strength respects strength. USA is the undisputed Space leader but they always found difficult to swallow the gargantuan cost of this type challenges costing approximately 100 billion US Dollar. (**One Billion USD = more than Rs 4,000 Crore**) thus other countries support is a must for their future ongoing programmes in the area of interplanetary Mars and Moon missions, solar satellite systems, space based solar power SBSP, ISS laboratory and technology for arresting global warming and climate change, pollution, drinking water, terrorism, defense platforms etc. It is clear that India and China have the means for cost reduction, motive and opportunity to become the technical leaders of 21st Century. Now Japanese Aero space Exploration Agency (JAXA) is working on ISS and astronauts are busy with the experiments in Japans KIBO lab and Hop laboratory in space.

After 2018 the full configuration will be completed and shuttle operations will start using falcon heavy reusable rockets by space x company. Using STS- Space Transportation System, USA and other space flaring nations can build kilometers of Solar Panel in space or constriction of huge global solar satellites for future energy sources from space.

The rocket cannot launch a football size laboratory and solar panels directly in to space at a stretch. Cosmonauts can go in to space and do the assembly in piece meals in Zero gravity. Rocket launching is incredible. Sometimes heavy loss can occur. Space X American rocket explodes at Cape Carnival launch pad two days ahead of launch on 01-09-2016. Space business is risky.

Final Configuration of ISS by the Year 2020

- Total Mass   - 500 Tones
- No. of Labs   - 7, Total pressurized volume- 1200 m3.
- Solar Power Supply   - 110 k w
- Low Earth Orbit- LEO   - Will vary between 335 km and 460 km altitude from earth surface.
- Assembly to complete in - 2020
- Nearly 17,000 NASA engineers and contractors across the country contribute to ISS and the space X private company for realizing the new generation rockets like Falcon heavy, space transportation system(STS) for the heavy lift to ISS are in progress. Falcon rocket is reusable with 9 semi cryo engines for several launch will reduce cost.

Infinite Space & Unlimited Excitements

- Permanent crew from 2012 onwards - 6 cosmonauts.
- Data transmission -Via TDRSS (Down-link - 43 MBPS).

In future the main competition in world space market may be with USA are (1) China (2) Russia (3) India (if the rulers of India having hard cooperation with Indian industry and global space industry in the world). India has been on Industrial growth. ISRO is outsourcing the work to Indian Industry for more job opportunity in industry sector. USA is planning to build a hotel in space by 2021 for tourism development in space.

India has built an information economy depend on satellite oriented space technology. Connecting Techno Park IT companies with qualified low paid and quality labor available in India to all world through satellite net work connectivity. This will help the country to an economical growth through software application. India is having approximately 80 crore youth between the age of 25 to 35. Value added leadership and budget allocation to industry can create job opportunity to extract work output for the development of India.

With this set up more economical growth can be achieved. India may get international partnership in business with developed countries like USA, Japan etc. for global space industry.

Indian Prime Minister Shri Narendramodi, American President Mr. Donald Trump and other world leaders may think an international partnership between India provided advanced technology is used for developments. Let us sacrifice our today so that our children can have a better tomorrow. The need to loft them efficiently and India's position near the equator, where the Earth's gravity is evenly distributed gives India a special incentive to consider building a geo synchronous space elevator. The other countries prefer to launch their satellite from India because of this *blessing and total cost effectiveness & precision for their missions. The key note for the Indian's style of space industry is to sell products in future to international market faster, better and cheaper.*

Stick with the goals of USA, UAE, Japan and India that improve the lives of common man. India launch EDUSAT a Satellite intended drench the subcontinent in science and technological education and another satellite meant for the youth all over the world as YOUTHSAT for International Youth Power connectivity satellite, 21st century Aero Space venture will be more global in view of unique space mission, technological complexities, high cost and high risk projects like Mars and Moon missions.

If we can design and develop rocket and satellites, India can manufacture civilian Aircraft and F-16 fighter jets. The modern fourth generation multi role fighter plane loaded with missiles also another option under the make in India initiative. This can be initiated as a public private partnership. As you are aware India can live without nuclear weapons. That is our dream, and it should be the dream of USA, UAE, Russia, Pakistan & China etc. to make earth a livable planet.

India always stands for the traditional values like, **Loka, Samastha, Sukino, Bhavanthoo, and loka skhemam maha maham** *are spiritual thoughts to make the earth a best livable planet. Space applications also a role to play and it must be used for peaceful purpose for the benefit of common man.*

GLOBAL SPACE INDUSTRY

Clean Energy supply in future may be the biggest challenge. One of the objectives is to give an idea to young scientist the future challenges in space industry. With this in mind, I would like to touch upon the future perspective and 21st Century Dimensions of space applications. There are lot of scientist sacrificed their life to make the earth a livable planet. The next generation also work hard to make the earth livable.

SOLAR POWER SATELLITES FUTURE POWER.

Solar power is abundant in space. Beaming the solar power back to Earth is not such a big problem. Once the power has been harvested by the solar arrays it is turned into micro-waves and sent down to a receiving station. In 1976, an experiment in the Californian desert showed that micro waves containing 30 kilowatts of energy could be beamed over one and a half kilo-meters with 82% efficiency. Hence, Collection stations across the Earth would gather the energy from the solar satellites, which could orbit the Earth in a fixed speed and distance as shown above.

With the present data, life is available only on Earth. Surprisingly few major environmental problems were foreseen. The engineers worried about microwave beams going away and cooking the local wildlife in the earth. The waste heat made by the collectors concerned them, and the land that would be needed for the receiving stations might be quite difficult to identify and obtain due to thick population and safety issues.

These challenges would be really very trivial compared to

the environmental problems that come with fossil fuel power stations liberating carbon dioxide and carbon monoxide, water vapors, methane etc these powerful greenhouse gases will hold the temperature even hotter. This rise in temperature due to global warming, will be the end of the life on surface of the earth. Cooling the earth down is a challenging task and very costly at present.

Microwave beams transporting the energy to Earth can be controlled and automatically switched off if they are not on the collector. The large receiver dishes, although a definite mark on the land, are not completely solid. Light and rain can get through them. In theory, the land under the collectors (which would be a paddy field) could still be used for farming. These receiving stations were only envisaged to cover a few square kilometers anyway, and they would take up no more land than conventional power stations. Even the waste heat from the collectors could be disposed of or even used as energy itself. Apart from pieces of equipment and old collectors that need replacing, the solar power satellites wouldn't produce much waste in the eco system in longer operations. It is still remain as science fiction. It can be meterialised through international co-operation like ISS (International Space Station).

THE INTERNATIONAL SPACE BASED SOLAR POWER STATION (ISBSP)

This **ISBSP** was a dream of late Dr A P J Abdul Kalam for future energy. Still it is a dream after 3 years of his death and it does not exist anywhere in the world. ISBSP can be similar to ISS (International Space Station) in configuration as explained in previous chapter. Hence Dr Kalam's dream will come to a reality when humanity will have acute shortage of electricity and drinking water. Necessity is the mother of inventions. 21 centaury needs, must work out for power without wire from infinite space.

Solar Power Sattellites for future energy

The burning fossil fuel like coal, diesel, petrol are polluting and warming our home planet at an unprecedented rate. The problem is becoming acute, with increasing population and standards of living across the planet. Fossil fuels are limited. Laid down over a hundred million years by the remains of sacrificial microbes and plants, the oil, coal and a great deal of natural gas hidden underground cannot keep supplying us with energy indefinitely to the coming generations.

Is there a better way of getting a clean energy that would allow us to maintain the Earth a Livable planet. National action plan is mandatory in Agricultural development, Electric Energy efficiency in transport vehicles, solar energy missions, sustainable habitat, drinking water, fresh air to breath, protection of eco systems, waste management, protection of rivers, green India, R&D on climate change and production of solar energy cost effective, water recourse management, Oceanography, new electric vehicles etc.

SUN is a nuclear fusion reactor

Solar power is abundant. SUN, pumps out energy which bathes the top of our planet's atmosphere with about 1,372 watts per square meter. This is about ten times more energy than the sunniest places on the Earth's surface. Out in space, free off clouds and the darkness of night, solar power is in continuous supply. The idea of building satellites to collect solar power in orbit around the Earth is viable, but very expensive at present.

A series of hearings was held by the US subcommittee on Aerospace Technology and National Needs, during which several engineers involved in designing solar power satellites, presented their ideas. When you read this book today, they are pervaded with an incredible sense of vision and foresight. This will also gave an impact on Gulf, oil price . In India solar power sector to shine brightly in 2025 on words. Public / Pvt sector Banks agreed to fund clean energy projects - capacity 76352 MW with an out lay of Rs 3.82 Lakhs Crore investment in power production.

The researchers considered solar panels 11 by 4 kilometers in size-vast constructions that would need a new class of heavy lift rocket to get them into space .Even today, with improved solar panels the devices would be very large and would need new space construction techniques. Costing several tens of billions of dollars in the year 2018, the system was never constructed because the ability to launch such huge machines was not pursued in the world so far.

One solution is to construct solar power satellites using resources on the moon using 3 D Printing, addictive manufacturing techniques. The lunar soil has plenty of silicon for making solar cells, and other metals and oxides that could be used to build the truss and the other parts of a working solar power satellite. The surface gravity of the Moon is just one sixth of the Earth's , so it would be much easier to tug the finished units out to orbit than it would be to launch it all from the surface of Earth. This plan depends on having a manufacturing base on the Moon a sufficiently complex to be able to build a solar power satellite, so this is not yet a practical solution for solving the problem of getting solar satellites into space.

THE FUTURE POWER WITHOUT WIRE FROM INFINITE SPACE

The SUN radiates about 10 trillion times the energy which human consumes across the world today. If we are able to extract even a small portion of this energy it would be sufficient to secure the energy demand of our future consumption. It is a clean energy; by producing it even in large scale will never pollute the environment. Space based solar power has many other advantages over traditional terrestrial or ground based solar panels we had seen along with the traffic **LED** lights in the road. First advantage in space based solar irradians is approximately 1.4 times efficiency than in ground level in earth surface.

Second advantage is in space collection time is 24 hours per day since SUN and satellite are stationary. In space it is not affected by weather condition like monsoon, wind, dust, corrosion etc. Solar panels in earth can only collect solar power 6 to 8 hours per day and affected by weather conditions. The power without wire is more efficient & environment friendly if it comes to reality in 21st Century.

The picture below shows a cosmonaut is doing assembly and integration work in space solar power for ISS. It has five major systems to address.

(1) Huge size power plant like five foot ball ground size to be assembled in space.

(2) Earth based collection system with safety protection like KSEB substations in earth.

(3) Medium of transmission wireless type or power without wire (micro wave and laser beams) from space to earth.

4) The solar plane which had traveled 42000 km across four continents,

two oceans and three seas. The final countdown was on 24 July 2016 from Cairo to Abu Dhabi (UAE). The zero fuel aircraft, solar impulse, has covered more than 42000km in its quest to become the first plane to circle the world using solely from the energy from the Sun is an example of power generation in Space.

5). Solar plane is not heavier than a car but with a wing span of a Boeing 747, the four engine battery powered air craft with 17000 solar cells embedded in its wings. The plane flied an average speed of 80 km per hour. This aircraft is encouraging to do more in solar power generation in Space.

DESALINATION OF SEA WATER - FUTURE POTABLE WATER

Another solar project in future required is desalination of sea water. Ocean top is blessed with huge solar energy. From the ocean we can produce drinking water and electricity with solar power. As you are aware in the earth 71% is water and 94% of global water

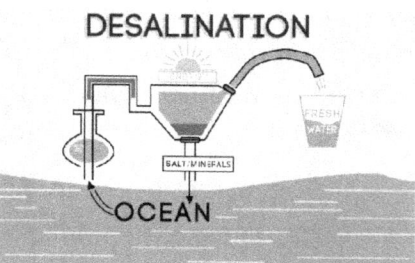

is in the ocean. With four percentage salt the ocean water is not potable. Top portion of the ocean water is hot by the sun light and bottom portion of water is cold due to depth. If we can collect water 1 km down in the ocean the average temperature difference between top and bottom is 21 degree centigrade. Desalination is possible making use of this thermo dynamics system. The hot water collected is allowed to pass through a vacuum chamber and the hot water started evaporating in low pressure inside the vacuum chamber. This water vapor thus obtained is again cooled by the cold water collected from bottom of the sea. The condensed water thus obtained is free from salt, potable and brought back to land for drinking. Solar water pump can be used for pumping cold water from sea bottom for cooling the vacuum chamber.

It must be a profitable projects and job opportunity of younger generation for the quest for future power and drinking water. They are going to face the shortage of potable water and electricity in future. Many agencies locally had started a project for protecting rivers in Kerala. It is also a good starting to make the planet livable with prosperity with the available resources of potable water.

Dr A P J Abdul Kalam had been advocating international co operation for the large scale space missions including space based solar satellites and drinking water to all. He always gave priority in interacting with youth for empowerment of youth power. He uses to emphasize the power of human mind embedded with good thoughts. His quote, **"Dream, Dream, Dream, dreams transforms in to thoughts and good thoughts result in good action."** This is very much true for innovations, to make the earth a livable planet. Terrorisms are also a product of bad thoughts in the mind of few human beings leading to destructive actions and an evil to the humanity throughout the world. Beautiful minds are an asset to the Nation. I would like to give few topics to 80 crore youth in India to Dream which may result to action later.

The youth in India must find out solutions through hard cooperation with R& D organization, because they may be affected much of the problems facing the humanity in years to come. Prevention is always better than cure. Life on earth and protection and sustainability of environment require many area of research and development like integrated study of atmosphere, global warming, accurate prediction of climate and weather forecasting, tsunami and earthquake forecasting, energy independence, water resources for the growing population, pollution in the air and water, waste management, recycling water, agriculture production, preventing adulteration in food and medicines, health care system, future power source to cop up increasing demand on electrical energy and fast track improvement of value added education system, Electric vehicles, design, e-waste recycling for building materials etc. required immediate solutions. They required integrated global approaches and synergy of all space faring nations and effective participation of young engineers and scientists.

The question you may ask to me is whether the international or national cooperation is consistent with the challenges in India and abroad for the next 50 years. Yes it is consistent only by hard cooperation of each nation contributing substantially in technology and resources in various projects. In this context I can tell you few examples of success.

One is the joint venture programme of India and Russia. The design, development and production of Brahmos missiles is an example. Only few people know that the world's first supersonic cruise missile made in India from Chaka, Thiruvanathapuram. Brahmose missiles in the defense sector project, leading to a business worth 1 Billion dollar export internationally is the hard cooperation of India and Russia. This is a success project under defense department (DRDO).

Another one is the Pan African e- net work programme of India and Africa. India had completed a project costing 120 million dollar of connecting 53 Pan African nations for providing tele-education, tele-medicine and e-governance for the development goals of Africa by United Nations. There are so many other examples in power generation sector like Indo-Russian friendship and cooperation of Thirunelveli koodankulam nuclear power plant project 1000 Mega Watts five numbers project. One 1000 Mega watts plant is operational in August 2016. Japan drinking water project in Kerala, metro rail project in Cochin etc where International cooperation result in dedicated funds and fixed time schedules leading to compatible finished products and spares made in India and abroad. Lithium-Ion, Battery large scale production is now started in India with joint venture with ISRO and BHEL for Electric Vehicles to roll in India by 2025.

The country can do indigenization of so many products like lithium ion battery for automobiles, solar panels for solar power in ground sector and on flexible Nano structure for solar satellite power generation and storage in space using Nano energy packs, transmission of electric energy from solar satellite in space to earth through micro wave and laser beam by wire les power etc. can trigger many young minds towards challenges which were impossible hitherto.

For a peaceful and safe world this paradigm shift in nations to work together is a noble idea. International friendship which is mutually beneficial may be a better solution for an enhanced quality of life in earth. A joint project by America and Australia called hypersonic international flight research experimentation are developing an engine that can fly in air at Mac 7. This hypersonic flight involves traveling at more than seven times the speed of sound. If they can achieve Mac 5, it will be a big revolution in the international flight sector to save time money and energy. The US company Boeing and German company DLR are targeting global air travel fast & access to space cost effective in future Space Tourism and Hotel in space.

In Japan, cities are connected with bullet trains so that people can reach everywhere in the country with les time. Here in Kazhakootam if all fast trains had given stop and introducing METRO trains people can reach to techno park from different place of Kerala on daily basis. They can reside in less cost area like Neyyatinkara, Parasala etc and earn more for house construction ,with more productivity in Techno park and city development programmes. The rulers has to promote the expertise of Technopark companies, Kudumbasree units, experts like Dr. E. Sreedharan etc. for the development of the State without any vested interest..

FUTURE DIRECTION FOR GLOBAL SPACE INDUSTRY

Space business is risky. Hence space industry also to take risk in large scale production. When the cost to access to space is brought down by one order or more, then this would open up avenues for new experiments and influence new players to come out with ideas. Some of the notable ones being discussed are space tourism providing an aircraft like solar impulse travel into space, experience the microgravity environment and return back to earth for a few tens of thousands of dollars.

The other potential space manufacturing is possible when ISS becomes routine and frequent which could throw opportunities for more travel from earth to space and back to earth. The above along with the ongoing scientific programs like Mars exploratory missions, Moon exploratory missions would provide the necessary input for long duration space travel to other planets.

One other important future potential, which is again being discussed, is the feasibility to generate electricity from space using huge solar panels of football field size and beam it back to earth safely as wire les power. Though, this is in a very preliminary stage due to cost, this is sure to become one important avenue with a very low cost. But, the primary requisite to make the space travel not only less costly but also more reliable and safer will remain as the key issues for the future generation for the benefit of mankind. Hypersonic flight fuel from atmosphere is a game changing technology and could revolutionise global air travel faster with scram jet expected to be completed in 2020. Space tourism can be established with low cost. ISRO flight tested scram jet engine for an air breathing propulsion system on 28-08-2016 joins the elite club for designing cost effective launch vehicles.

Summary of Solar Power Generation And Transmission to Earth
Why Solar Power?
Necessity, Advantages & Disadvantages

- Fast depleting conventional power sources like diesel, petrol, coal, etc. pollution and high cost.

- Increasing energy demand due to the world population increase to thousand crore in the year 2050. At present population is 750 crore. 33% increase required in future power generation for the demand and supply for electric vehicles to roll.

- Pollution from conventional energy production techniques and associated global warming and other environmental problems like climate change shortage of food and drinking water.

- Threat from nuclear power plants, radiation, lack of new technology, raw materials availability etc.

- Increasing use and high costs of electricity by the demand.

- Hence a renewable source like solar power is essential in future.

INTRODUCTION (Training the human brain to think)

- Renewable and clean energy is an impending necessity.

- Space –based solar power (SBSP) is the concept of collecting solar power in space for use on earth. Now it is a dream project.

   SBSP would differ from current solar collection methods in that the means used to collect energy would reside on an orbiting satellite instead of on Earth's surface affected by earth rotation, cloud, dust, wind, rain, corrosion, cyclone etc.

ADVANTAGES, IF POWER IS FROM INFINITE SPACE

- To make solar energy affordable, we must put the collectors in space. Where the sunshine 24 hours and the intensity of sunlight is about 1372 W/m2. Intensity is 40% greater than on Earth.

- The best location is geostationary orbit (GSO, 35,800 km above the equator), where a satellite remains fixed relative to terrestrial sites.

- The principal components of a power satellite are a large solar array and a microwave transmitter that beams power to an Earth-based receiver called a retina, where it is converted to standard AC supply for distribution in earth.

- The continuous, intense sunlight in GSO means that no energy storage is needed, and that the solar array is a factor of eight times smaller than a similar terrestrial array with the same average output (storage battery is not used in the space based solar power).

- The benign operating environment, in vacuum and free fall, permits high solar concentration without complex sun-tracking mechanisms and avoids maintenance problems caused by wind, dust, rain, snow or hail corrosion cyclone etc.

- Re- directable power transmission: A collecting satellite could possibly direct power on demand to different geographical locations based on maximum demand and peak load power need in earth.

BASIC PRINCIPLE

ISBSP, Space Based Solar Power essentially consists of four elements to think:

- A means of collecting solar power in space, for example via solar cells or a heat engine. Huge structures of solar panels 10 km X 5 km size.

- A means of transmitting power to earth, for example via microwave or laser. Wireless power. Importance to safety in transmission to earth is a challenging task due to population.

- A means of receiving power on earth, for example via a microwave antenna (retina, eg: radio frequency retina, optical retina).
- A laser pilot beam guides the microwave power transmission to a retina (rectifying antenna).

COST ASPECT AND CHOICE OF LAUNCHING ROCKET APPROXIMATE ESTIMATE FOR WEIGHT & SIZE

- One problem for the ISBSP concept is the cost of space launches and the amount of material that would need to be launched to space. Now Falcon Heavy can do this by Space X Company, USA
- Currently the cost of access to space $25,000 per kilogram for the space Craft, with payload fraction of 1.5% and is reusable about 10-100 times per airframe, with major refurbishment following each flight.
- The Solar Array Mass ($\approx$ 100w/kg works out to be 10000 Tonnes, requiring 400 flights costing $250 Billion with the current technology, to generate 1 Giga watt electricity from space.
- To achieve a cost of not more than a few 100s of dollars per kg in LEO, a Space transportation system has to be reusable at least 100-1,000 times (with significant refurbishment between flights), and to have a payload fraction at least 5-10 times that of the space shuttle; namely, Falcon Heavy. Private company name Space X. USA owned by Dr. Elon Musk may give hope for this project.
- Such high payload fractions are attainable only when the vehicle carriers no liquid oxygen board at take off, but collect and liquefy oxygen while climbing to orbit in hypersonic flight regime.
- Much of the material launched need not be delivered to its eventual orbit immediately, which raises the possibility, that high efficiency (but slower) engines could move SPS material from LEO to GEO at an acceptable cost. Examples include Electric or Ion Thrusters or Nuclear propulsion for orbit raising and station keeping activities.

PRESENT COST OF POWER PRODUCTION ROUGH ESTIMATE

- The present cost of Multi-Junction Cells ≈ 300 $/watt, Hence 1GW it costs 300B$; with this cost, it would take 171 years to recover the investment, without even accounting the cost of launch It can be reduced with new technology in future.

- A Quantum improvement would be required, to bring down production cost of high efficiency Solar cells and heavy lift rockets using global space industry production units.

- R & D to improve amorphous cells based flexible solar arrays and light weight solar structure development and Air breathing rockets reusable type is needed.

NEED FOR INTERNATIONAL CO-OPERATION TO REDUCE COST? (India can use youth power to industry)

- Rapid progress can be made more effectively if the interested countries work together to establish an enduring, firmly integrated consortium of share holders, in order to further the development if international space-based solar power stations and various required enabling technologies such as safe, affordable and reliable access to space, wireless power transmission, advanced in space operations, with others jointly.

BASIC DESIGN SPECIFICATIONS FOR DISCUSSION IN R&D FORUMS.

- Targeted Power Generation: 1GW. Solar array area required: 2 Sq Km (250 W/M2). Solar array Mass: 10000 Tons. Ground Collection Antenna: 100 Sq Km. Connection to existing Power Grid related plants sub stations etc..

SCOPE OF WORK

- Establishment the short term and long term goals for ISBSP energy requirements capacity to pay for the technology and energy production.

- Detailed study of the market survey its characteristics and formulation of laws. Study of the design and technology of the ISBSP by expert teams (International co-operation).
  - ✓ The systems Architecture for the first ISBSP missions.
  - ✓ The technologies required and their sources internationally.
  - ✓ Cost and time for development of a Technology Demonstrator.
- Scope of utilizing existing/near future launchers for space transportation and construction requirements. Approval needed from world governments and international bodies and the means to get them from all developed countries like USA, Japan, China to the developing countries like India etc.

HOW TO GO FORWARD IN PHASES?.

- The steps forward proposed here are to be taken up in a self-renewing manner in several phases:
- **Phase 1:** *An International PRELIMINARY Pre-Feasibility Study team with all space agencies in the world for a small ISBSP design.*
- **Phase 2:** *A Detailed International Feasibility Study, coupled with a range of targeted Engineering Demonstrations on Ground and Technology Experiments in space. Team from NASA, ESA, ISRO, China etc for a proto type ISBSP design.*
- **Phase 3:** *Development and Deployment of one or more initial Space Solar Power Pilot Plants of one or more initial space solar power pilot plants, capable of delivering meaningful power to terrestrial markets from prototype plant assembled in earth.*
- **Phase 4:** *Space Solar Power (SSP) and supporting infrastructure. Industrialization (self-sustaining & self-renewing far into the future). Faster, cheaper and better ISBSP production.*

TECHNOLOGY ELEMENTS TO BE DEVELOPED

(1) Development of solar panel arrays with high conversion efficiency at low cost. (2) Development of suitable launch system for outer space transportation of solar panel & Concentrator. (3)Heat Engine deployment and maintenance.(4) Development of microwave, Laser transmission system.(5) Development of receiver and distribution system on earth.

CONCERN AREAS

Sun emits about 70 millions watts per metre square energy and earth receives 1372 w/m^2 solar energy. Transmitting large quantity energy from orbit to Earth's surface required safety and reliability due to population. Since wires extending from Earth's surface to an orbiting satellite are neither practical nor feasible with current technology, ISBSP designs generally include the use of some manner of wireless power transmission. The collecting satellite would convert solar energy into electrical on board, powering a microwave transmitter or laser emitter, and focus its beam toward a collector on the Earth's surface. Radiation, space debris and micro meteoroid damage could also become concerns for ISBSP.

DISADVANTAGES EXPECTED

- The space environment is hostile; panels suffer about 10 times the degradation they would on Earth. System lifetimes on the order of a decade would be expected, which makes it difficult to produce enough power to be economical in long time working.

- Space debris is a major hazard to large objects in space, and SBSP systems have been singled out as a particularly hazardous activity.

- The broadcast frequency of the microwave downlink (if used) would require isolating the ISBSP systems away from other satellites. GEO space is already well used and it is considered unlikely the ITU would allow an ISBSP to be launched to the commercial satellite orbits in GSO.

- Only about half the power generated by the ISBSP would be delivered to the grid. Once all losses are factored in. These losses are on the same order as modern fossil fuel plants using coal and naphtha now working in earth for producing electricity.

CHOICE OF RETINA TO CONVERT LIGHT IN TO ELECTRICITY

Radio frequency retina: an antenna and a demodulating diode. Optical retina: scaled down to nanotechnology proportions could be used to convert light into electricity. Each satellite will deliver 2 GW to the utility grid, an output similar to a large nuclear plant. There is room in GSO for thousands of them. The microwave flux in the power beam is insufficient to harm aircraft or birds. The retina area is a factor of nine times smaller than the terrestrial solar farm that it replaces; it can be located close to the intended load centre; and the structure shields the ground underneath from microwaves but is largely transparent to sunlight, so that land can be used for agriculture or any other work.

This chapter is only for training the youth brain to think for future power. ISBSP is not existing in the world. This may throw light on the possible application of ISBSP. The contents are like a science fiction or fantasy. I hope that ISBSP will come to reality, when there is acute shortage for future power due to increasing population in the world and introduction of Electric Vehicles. Yesterday's dream, tomorrow's reality. ISBSP and space industry are dream projects. On reading it, younger readers will want to be a part of it for the adventures like man mission to moon and Mars in 2030, human habitat in 2050. When we look at the magnitude of universe, we cannot restrict it to our solar system alone for future power. A fusion reactor also a better option for future power. Hydrogen atom converts to Helium atom constantly in the fusion reactor for the production of electricity using the thermal power.

MISSION TO MOON & MARS

By the year 2050 when the World population touches one thousand crore it is necessary to find out energy and drinking water for future generation. The era of wood and bio-mass is almost nearing its end. The age of oil and natural gases would soon be over even within the next few decades for the existing 740 crore world population. Every day oil price is going up the world energy forum has predicted that fossil based oil like petrol and diesel Coal and gas reserve will last to another five to ten decades. Implementing Nuclear power and its deal is creating a lot of debate in the government level for the alternate energy for future generation. Water for future generation is also a big problem for the increasing population. Water everywhere, but a single drop is not available for drinking. This may be the situation due to pollution in rivers.

    More than 71 percentage of Earth surface is having water and only one percentage is available as fresh water for drinking purpose. Minerals and metals for the future generation are not sufficient in the Earth. Because of the increasing population 33 % more resources required additionally, man has to go to other planets in search of resources to live in prosperity. The youth in the world has to make strategic plan for this habitat. They have to struggle for the existence in the earth.

    "THE MOON & MARS" with the history of the early solar system etched on it beckons on mankind from time to discover its secrets. Understanding the moon and mars will provide a pathway to unreveal the early evolution of earth. Through the ages, the Moon, our closest visible celestial body has aroused curiosity in our mind much more lead to scientific study of the Moon, driven by human desire and quest for knowledge. This help the moon got a boost with the advent of the space age and the decades of sixties and seventies manned missions to moon. This mission reveals about the origin and evolution of the moon and its place as a link to understand the early history of mother earth.

However, new questions about lunar evolution also emerged and new possibilities of using the solar system and beyond were formulated. Moon again became the prime target for exploring. All the major space faring nations of the world started planning missions to expect potential base for space exploration. ISRO also planning the Chandrayan 2 mission with a moving rover in 2018.

Characteristics of the Moon

- Age 4.5 billion years age. Distance from earth 3,84,400 km.
- Diameter 3,476 km; (¼ size of earth's diameter).
- Mass 7.35×10^{22} kg (1% mass of earth)
- Surface area 3.79×10^7 km^2 (7% area of earth)
- Mean density 3.35 gcm^{-3}; 5.52 gcm^{-3} for earth
- Gravity 1.62 m/s^2 (17% of earth). Escape velocity 2.38 km/s.
- Other names Chadra (Sanskrit), Luna (Roman), Selene (Greek), Chang'e (Chinese)- Chandryayan – ISRO Mission to Moon.
- Special features. Abundance of Helium-3 - Absence of Iron.
- Low Gravity on the Moon. Perfect Vacuum on the Moon.
- Temperature extremes on the Moon. 110^0C during day time.
- Moon rotation about axis and earth. Because of this we can see one face of the moon only from earth. Moon will soon become a telecommunication hub for interplanetary communications.
- Orbital velocity 1.68 km/s. Temperature (on near side) 110C On far side very chilled minus 180^0C. Syndic period 29 days.

HUMAN MISSION TO MARS, PERSPECTIVE & CHALLENGES.

In 2018 the world population with 740 crore people are not happy with their need for energy, food, potable water, minerals & metals etc. In 2050 the world population will be 1000 crore. We cannot even imagine the unhappiness of people after 32 years, if 33% increase is not made in their need for energy, food, potable water, minerals & metals etc. In that situations people may think to go to Mars or Moon for a happy and prosperous living. No competition for all resources like land, sand, rock etc. It will be very easy to make habitat with the help of real estate people and space scientists at Mars.

Mars the red planet & moon the visible celestial body from earth are our close neighbours. These two have always attracted by mankind on earth since the dawn of civilization. Seasonal variation, moving storm with dust, the presence of water in the form of frozen ice, evidence of flowing water like river may harbour living organisms has further raised chances on Mars for human habitat in 2050.

Mars may be the next destination for man after Armstrong placed his first step in the moon almost half of century ago. Space scientist had an opinion that Mars is not that simple as the moon due to various reasons stated below. With the present technology human habitat in Mars is not possible. The reasons are as follows.

The travelling time to reach Mars from earth is the basic hurdle. Moon is just half a million kilometres away and to Mars is

about 150 million kilometre away from earth. Moon can be reached in few days. Mars journey will take 8 to10 months. The passengers to Mars has to stay minimum 500 days in the rocket with food oxygen water etc taken on board from earth and in between no stoppage for food and toilets. No hotels in Mars and on the way to Mars are another hurdle. This can create physiological and psychological implication in addition to solar radiation and prolonged weightlessness in the deep space. Kerala people definitely get bored in the journey without traffic blokes and processions and harthal by political parties on the way to Mars. Youngsters can enjoy driving the rocket with supersonic jet speed or more, one Lakhs km/hr? Without stopping in between Earth and Mars (extremely smooth journey).

The above are the problems of passengers. Now we discuss the challenges of space scientist for developing the rocket. Considering the longer duration with five crews the injection mass of rocket to low earth orbit is 1000 metric tons. Soft landing to Mars is another challenge and again an ascent from Mars for the return journey is also required on board communication. NASA had developed real time communication systems for international space station (ISS) & Moon landing. Android mobile phones are now working with this technology developed by NASA for Moon landing. For Mars, because of distance 20 minutes delay for immediate consultation with earth is also a challenge. This demand more autonomy in deep space signal processing on board. For long duration propulsion solar electric propulsion is another option under active consideration. The landing site selection must be near to water and other minerals. Robotic work force is to be brought to Mars for construction work. They are controlled and guided by crew in orbit around Mars. Return to earth is also a serious problem at present.

NASA proposes to test out many of this idea and they suggest establishing a habitat first in Moon. Moon can also use as the fuelling hub of rockets bound toMars. Meanwhile a Dutch company named "Mars One" had advertised for workmen interested to go one way trip to Mars in 2030 to start human colony there. Mars foundation promises to launch two people a man and women on a 501 day trip to Mars and back in 2023. Those who are depressed to live in earth can make use of this opportunity to go to Mars and enjoy.

NASA is forging ahead with the space launch system (SLS) Falcon Heavy and the Orion space craft to send human into deep space and propose "Human orbiting Mars" towards establishing a sustainable Mars exploration programme. This is targeted to take place in 2030 which will be followed by human habitat on Mars in 2050. With international cooperation NASA and other space flaring nations can construct villas in 2053 when earth population touching 1000 core. One has to wait and see how things evolve. However, when we look at the magnitude of Universe, we cannot restrict it to our solar system alone for exploration. Then landing on Mars will again qualify as a small step. But how? This question I am asking to young scientist. Dr. Elone Musk of space X, USA is having strategic plan for Mars Mission

Need aspect for reducing the cost is mandatory for space applications and inter planetory missions..

Dr. Elon Musk, Space X Company, U.S.A is a living legend. He can enable the full commercial potential of space and expansion of space research and exploration of Universe. To accomplish this goal, access to space must become orders of magnitude more affordable.

We have exhausted the possibilities for dramatic improvement via available technology, and must invent new technologies across a broad front to overcome the obstacles that lie across the highway to space. With sustained effort, starting from converged requirements and guided by logic, we can achieve our goal through international cooperation and friendship through world youth connectivity.

On the commercial front, taking advantage of the huge market for satellite building and launch, there is a surge of launch vehicle development activity occurring around the world as private contractors and government organizations strive to gain a share of the dynamic and rapidly evolving commercial satellite market and interplanetary missions to Mars. More than a dozen major launch vehicle programs are planning all-new families of vehicles or developing more capable derivatives of existing models, while a comparable number of smaller companies are attempting dramatic launch cost breakthroughs in specific market niches with innovative booster designs and air breathing propulsion, semi cryogenics etc.

In the near-term, the cumulative effect of this widespread development in launch vehicle heavy lift, according to industry observers, especially as the commercial satellite market is flourishing. The overall long term outlook, however,  *remains relatively bright, in the estimation of many observers in the space industry worldwide.*

Reusable launch vehicles is a better option for cost reduction

The Space Shuttle is the first generation reusable launch system and represents only a part of what is possible in space.

Various nations have been exploring the design and development of RLVs for several decades. During the mid-1990s, however, several small "start up" companies in US, staffed with entrepreneurs, engineers and other technical personnel with years of experience in the aerospace and launch vehicle industries, proposed to develop commercial reusable launch vehicles. While many of these vehicles are being built with the expectation that there will be significant demand for launches of communication and remote sensing satellites to low Earth orbit (LEO), some hope to serve other new markets such as space station resupply and commercial microgravity missions and ISBSP etc.

Some RLV operators also hope to serve the market for geosynchronous Earth orbit (GEO) launches as well. In addition, over a dozen RLV designs are proposed specifically to help the market for space tourism. Also during the mid-2018, NASA began development of test bed vehicles that would prove technologies and operations concepts for next generation RLVs.

Falcon 9 with marline engines, space launch system (SLS), Orion & Falcon heavy rockets are reusable type by NASA. For space flaring nation's sky is the limit for exploration; however we will use the present 'orbits' for value addition to space explorations. These orbits are comnely used for international space market for space applications in the years to come. "Loka Samastha Sukhino Bhavanthu"-"the earth is the cradle of human kind, but one cannot live in the cradle for ever".

CHAPTER 7

GLOBAL WARMING & CLIMATE CHANGE
PERSPECTIVE & CHALLENGES

Why we study global warming? Why should we invest considerable Time, Money & Energy to arrest the phenomena of climate change?

The reason is both theoretical and practical. The theoretical reasons are **SUN IS HEATING UP & EARTH OZONE PROTECTION LEYAR IS DEPLETIING**. The practical reason is **GREEN HOUSE EFFECT**. The after effects are global warming and climate change. India is highly vulnerable to these phenomena. This will lead to shortage of sea food, agriculture production, drinking water, survival of species like honey bees. Birds. Butterflies, which helped plants for pollination to produce crops. Adverse for the survival of mankind in earth etc. As climate changes worsen, some coast lands could go underwater and other regions could suffer extreme heat and droughts causing massive human suffering. Floods and cyclone can also occur sometimes end up with unpredictable disasters globally.

SUN IS HEATING UP?

Now we will study the Physics of Global Warming. The amount of radiation pouring out of the sun towards earth is growing. This will heat the surface of earth and ocean to the point that the evaporation rate increase in oceans and dry rivers. The water vapor thus produced will increase the humidity making global warming more difficult to manage in coastal areas. Unpredictable cyclone wind will occur in costal area due to Global Warming and climate change.

As shown in the figure below at the core of the sun the high temperature and pressure convert Hydrogen in to Helium. For every tone of material the sun converts, it shrinks a bit making the sun denser and little hotter. Our mother Earth is 100 times smaller than Sun and orbits at a distance 93 million miles. However she is getting her share of light and heat through radiation.

History says that in some 4.6 Billion years ago the beginning of earth may be with abundance of water, Carbone Dioxide and Nitrogen. Then Chemical Era started with water condensed in to ocean and Carbone Dioxide dissolved to form carbonate. Some 600million years ago Biological Era stated with Photosynthesis process by forest increases the Oxygen content in the air to the present level. Some 100 million years ago Geological Era started and continents drift resulting changes in climate/ weather pattern making Earth livable. Now we are in Anthropogenic Era.

Physics of Global Warming

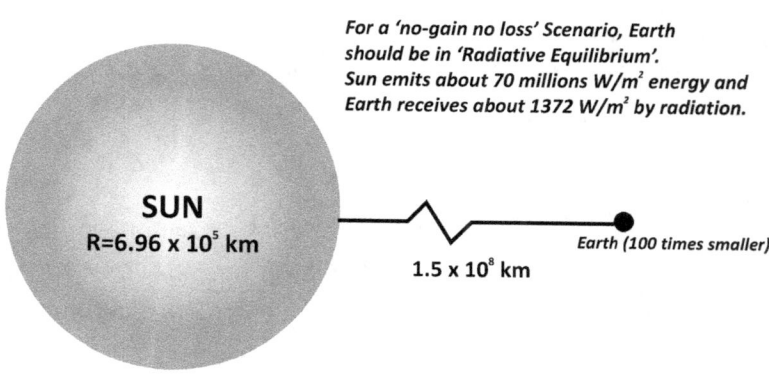

For a 'no-gain no loss' Scenario, Earth should be in 'Radiative Equilibrium'. Sun emits about 70 millions W/m^2 energy and Earth receives about 1372 W/m^2 by radiation.

SUN
R=6.96 x 10^5 km

1.5 x 10^8 km

Earth (100 times smaller)

WHAT IS GREEN HOUSE EFFECT?

In the history of atmospheric evolution, now here is a dangerous and a very powerful driver the human beings. The HUMAN GREEDS changes the composition of the atmosphere more rapidly, than ever before by introducing Industrial Era- producing Carbon Dioxide (CO_2), plastic, methane etc.

Due to this the global atmospheric concentration is finding difficult to manage the influence of three well mixed green house gases like Carbon Dioxide , Methane, Nitrous Oxide and water vapors. Emission of this green house gases alarmingly is the major reason of green house effect. Thousand Metric tons of Carbon Dioxide were emitted daily to atmosphere from fossil –fuel burning, by automobiles cement production, power generation using coal etc. Now we will study, what is green house effect?

It is very obvious that, the above well mixed green house gas, is the main culprit for greenhouse warming. These Carbon Dioxide layers with other gas are acting as a shield on the top of the earth. The heat radiated by the Sun and burning of fossil fuels like coal, diesel, petrol etc is not going out from the Earth's surface. This situation is more complex in cooling the earth down. This hot condition on earth is called green house effect.

For example, a car parking under the hot sun with closed windows, will feel too hot inside. In this example the closed environment inside the car is creating the warmness. This heat is always more than outside temperature. In the case of Earth CO_2 shielding phenomenon is called Greenhouse effect. Green house effect is neither a science fiction nor fantasy. We can experience it.

The world population is going to struggle with this effect. In India most of the cities are suffering from pollution. During summer season day time temperature is more than 45° Centigrade will make the life miserable. The quality of air in cities has now crossed the threshold pollution level due to burning of plastics. Novel approaches in climate engineering have been suggested to bring temperature down. Increasing green house gas amount increases the green house warming. The practical solutions are the following.

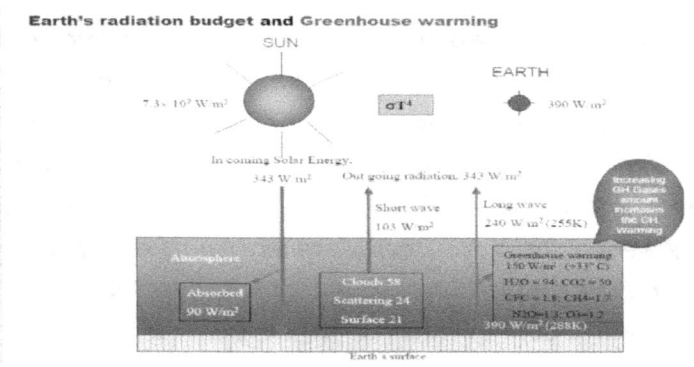

Human and other living things on Earth have evolved to deal with Earth's unique environment with an atmosphere. We have a strong backbone, along with muscle and connective tissue, to support ourselves against the pull of gravity. On Earth, the ozone layer and the magnetosphere protect us from solar radiation and charged particles. The increased rate of Carbon Dioxide gas mixed with other Greenhouse gas created damage in the Ozone layer and allowing solar radiation and charged particle to reach on earth surface .It is mainly manmade problems due to the large production of Carbon Dioxide gas. With the extreme hot the water in rivers and oceans will evaporate, leading to rise in humidity levels. A spike in humidity, especially in the costal climate, can lead to sweating, uneasiness, sun burn and heat stroke. Cyclonic wind can also occure in the sea shore, volute to form deep depressions in oceans leads to heavy rain fall.

SOLUTIONS

Carbone emissions reduction approach to fighting global warming is one method. Civil societies especially youth have to come up with a broader collective approach to protect the environment. Each colony or resident associations should be encouraged to have some green cover by planting Trees or gardening and developing a system for waste management. Social media can help in creating awareness and increase peoples participation in planting Trees. The trees can absorbs carbon Dioxide, by photosynthesis, bring down air pollution. Trees can store water, and atmospheric nitrogen, which enhanced fertility. Trees reduce sound pollution and provide food and habitat for other life forms like honeybee, birds, butterflies etc.

Promote an integrated approach to environment friendly living and managing home as shown in the flow chart, can be initiated individually, to cop up global warming. This chart is for individuals to follow mandatory for cooling the earth down.

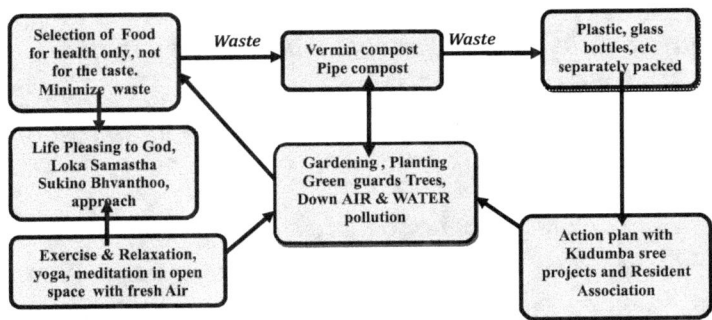

CONTROL CARBON DIOXIDE or SUN LIGHT or BOTH?

Most climate engineering practical efforts can be divided into two categories which address respectively, the management of the carbon and the management of sunlight. The figure below shows the Earth radiation processes for controlling carbon Dioxide and sunlight or both.

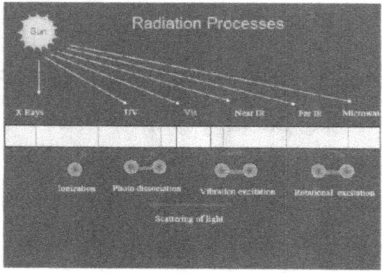

The first category is directed towards removing green house gas from the atmosphere. A prominent example is Carbon Capture and Storage (CCS), where some of the carbon dioxide being emitted by coal-fired power station is recaptured by physically sucking it in and transporting it elsewhere to be sequestered underground. The first 115 MW CCS retrofitted coal power plant commenced operation at Boundary Dam in Canada in 2014.

The CO2 captured there is transported and pumped into the nearby oil field for enhanced oil recovery. This has reduced its CO2 emission by one million tone each year. Studies are on in the U.K and other nation on the feasibility of similar installations there.

Another method for removing CO2 from the atmosphere is to increase forest cover as plants will absorb some of the unwanted CO2. Increased forestation is part on Indians' strategy for reducing CO2. In cities burning of plastics and wastes creating smokes enriched with Carbon Dioxides. In managing home an integrated approach to environment friendly living with reduced cost of living is possible in each family.

More ambitious, but also more costly and complicated, is the second category of climate engineering: Solar Radiation Management (SRM). Here the plan is to, reduce global warming by cutting down the heat absorbed by our planet from the sun. Among the technique being considered are marine cloud brightening, cirrus cloud manipulation and stratospheric aerosol injection (SAI). SAI, the boldest and also the most risky of climate engineering interventions, involve spraying into the atmosphere fine, light – colored particle designed to reflect back part of the solar radiation before it reaches and warms the earth. SAI proponents claim that this could bring down the global temperature by as much as five degree centigrade a substantial amount in the climate change context. SAI research is still theoretical and costly at present.

Much preliminary research has already been done on this technique and reviewed in major journals and News Papers. The optimal gases for injection, such as Sulfur Dioxide (SO2) can be produced in abundance. Furthermore, just a few aero planes specially redesigned for the purpose may suffice for injecting the required aerosol into the atmosphere. There are also precedents from nature. The 1991 volcanic eruption of Mount Pinatubo in the Philippines injected 20 mega tones of Sulfur Dioxide (SO2) into the stratosphere, cooling the globe significantly for a couple of years.

FLASHPOINTS OF THE FUTURE OF GLOBAL WARMING

Under such pressure, and in the absence of international regulatory regimes, the affected nation, even a small developing one, may well resort to using whatever SAI (Stratospheric Aerosol Injection) technology is available by them. Using a few aero planes to inject aerosol to bring the temperature down by one degree may cost billion dollars. U.S.A is not willing to contribute the carbon credit fund to developing countries as per Paris treaty. India is supporting climate fund and 'Swacha Bharath Mission' started in a big way.

R & D is in progress to generate no Carbone energy like solar power. Watching the Earth from space seems a rather passive environmental benefit of exploring space and understanding the changes occurred by global warming. We are improving our ability to fine resources, use them wisely and watch the impact on the Earth, but all this is just a product of International collaboration and global responsibility to arrest the change with cost effective methods. As temperature rise, complex life forms, will make life on Earth is not comfortable for the next generations. Prevention is always better than cure for cooling the earth down.

HUMAN STRESS MANAGEMENT

Scientists working in the area of rocketry also influence stress. There is no partial success in this area of burning large quantity, high energy propellant. Space will never forgive to any agency and hence it is a risky business in the world.

Stress is essentially a body's reaction to a threatening impulse to our mind. The longer you hold, heavier it becomes. Learn to put it down and pick it up when you need, is stress management.

Human mind is a very good fertilizer, anything we plant inside like stress, love, hate, fear, hope, revenge, jealousy; possessiveness, definitely it grows and bears fruit. We have to decide what to harvest. With the right attitude you can inspire and motivate yourself to do much more greater things for a stress free life. Stress is another factor which can induce the body to store more fat and develop blokes resulting to cardiac problems like heart attack and psychosomatic diseases, memory loss etc.

The psychological stress such as head ache, anxiety, anger, depression, emotional outburst, weight gain etc will make self damage and others life miserable. Feeling of isolation affects the digestion and sleep pattern and do a lot of damage to endothelium, inner most layer of arterial walls and stomach upset and pain, migraine, vomiting etc. Stress prevents the liver from secreting H D L, a protective element of human body called as good cholesterol.

How to manage stress? There is always a solution to every problem. Solutions may be neat, plausible, and sometime wrong. No one will make a lock without a key to open. Similarly God will not give problems without a solution. We cannot kill stress. But stress can be managed. Life is an echo. What you send out will come back to you as action and reaction. Your thoughts are the generator of stress. Stress is an attitude problem. The attitude  shows a person's altitude. So this paper is discussing about the attitude changes rather than medical treatment and medications. Yoga and meditation experts proved that the human body and mind have to work in harmony for developing health and happiness. .Peace and health comes when physical, mental and spiritual energies find harmony within our body to work more than Billion cells. The stress will never allow you to be healthy and a joyful person. This phenomena result in to stress induced diseases and frequent visit to hospitals and making the life miserable fallowed with behavioral issues.

Today's life style is, too much to do, too little time, too little energy. This life style will offer threatening impulses to mind and body to react accordingly. We can imagine as body is the hardware and mind is the software. The software gets corrupted due to stress bugs and whole body functions are affected. The person becomes abnormal in behavior. Now we will see a case study about, how the stress is developing spontaneously and managed by ancestors early days.

Think of us being here four hundred years ago in Thiruvananthapuram. At that time most of the places are 'caud' (forest). Even now see the name of places, Thycaud, Vazhuthacaud, Manacaud, Neduncaud, Nedumancaud etc. Obviously there won't be any corporation or town around us, and population and garbage also very less. We would be in a forest with full of wild animals. As soon as we saw a lion in front of us in Thycaud we will be stressed. Due to fear and emotional stress our breath increases. We have two options. Fight or Flight. Both this option required a higher blood pressure .The brain starts pumping adrenalin into the blood stream. More blood is pumped to legs. We fly away from the attack of lion and life is saved. From our experience we know that fighting with the lion is not a good choice. We use the blood pressure for energizing the legs to run and manage the stress due to fear. Stress is managed well for the escape from the lion. This shows that a minimum stress is required to do something, otherwise we will do nothing. But if the stress crosses the breakdown level it must be managed.

In today's busy life style there is no forest or wild animals but politicians, harthal, liquor, in law problems in micro family, different type fever, afraid of mosquitoes, garbage in public places, traffic blocks, adulteration in food and medicines, Global warming, climate change and so many unforeseen issues in family and work places etc. We cannot do either- fight or fly away from the state or the house and family. Fight with the in laws, boss in the organization and political parties in the state are also not practical always since you are alone and no support is possible by others. After all our small life is not spent for quarrels. Things are much more pathetic those who are working in Gulf countries & Techno park for struggling for existence in the profession and managing a happy family life abroad. The stress created by the above people is accumulating in human mind and not able to vent by running out from the problems. Stress developed is holding for a longer period and damaging the systems, if we do not know how to manage stress.

UTILITY OF STRESS

Developing stress is not really a person's problem as long as you know how to manage the stress. In fact stress up to a point is useful for two reasons. Firstly, it makes you productive. If you have no stress, you may produce nothing in time. At a certain cases this can call as useful stress. Any additional stress brings down your productivity very sharply. Instead of working, you start worrying about your work, head ache, sleep les night, drug addiction, cursing about our fate etc. In this point the stress has to be managed. Another aspect of stress is that, if you get accustomed to a certain level of stress, you are uncomfortable either above or below that level.

Now we will see a case study. Some of the house wife had fairly hard life in bringing happiness in family, managing the kitchen and office, needs of other family members and satisfying husband's egos in law problems etc. since all women are working women either in kitchen or in office or both. Irrespective of age stress can go up to break down level daily. House wife normally will not get appreciation or respect from family and no retirement from kitchen and cooking the most difficult job. This situation will continue up to the age of 60 for women. When the children got married and went abroad, husband retired , son got married no cooking & having daily eating only bread and banana, viewing T V. The household tasks suddenly got reduced. The housewife started judging others, no time to love other family members. Always having complaints like knee pain, Thyroid problems, back ache, depression, sugar, cholesterol, cardiac issues, constipation and finding fault with others even the Prime Minister, corporation Mayor for garbage management etc. There is suddenly no work other than the above. Hence no stress. Then the daughter comes for her delivery. Immediately all ailments vanished. The grandmother was very active, after the grandchild was born. For the next four months, she went through a lot of hardships but she was as healthy as ever before because of busy schedules of the kid. After four months, the daughter goes away with the grandchild and all the ailments return to the grandmother. By seeing this enthusiasm, the daughter and son in law are very happy to take her mother for child care at the cost of an air ticket to an extreme climate country. But the Grandfather in the age of sixty plus is alone in house looking after the big house and other properties. No demand for him since he is not productive and having some medical problems and the treatment is very costly abroad . Retired officials are the victims of this type stress. Longevity will give more stress.

Thus stress increased above a certain level can also be a problem especially when you are getting very old. Please remember diet control and exercise extent your old age and not the youthfulness. The old age stress is a major problem because of the longevity of life extended up to the age 95 plus. So in managing stress, we don't always increase stress, but keep stress within the limits to make the stress both comfortable and productive as described above. Do not idle and do some work to keep you under minimum stress and productive. Now we will discus-

TYPE OF STRESS & HOW TO MANAGE EACH STRESS.

Mainly there are six type of stress to manage.

1. **Mental stress**
2. **Psychological Stress**
3. **Family Stress**
4. **Career Stress**
5. **Economic Stress**
6. **Social Stress**

All these above six stresses, if you fail to keep within the limit, people tend to exhibit life style deceases such as blood pressure, hyper tension, stroke, coronary artery disease, kidney problems, etc .Now youngsters below 30 years old are exhibiting these disease. Due to break down stress people are not able to control anger and tongue. Human child will take two years to learn to speak.   When we grown up, it will take a life time to learn what not to speak to others. Individuality and self esteem is very important. It is our duty to protect it at any cost. Do not compare with others. It can reduce mental stress. A positive mind finds opportunity in everything. A negative mind find fault in everything. With this theory we can find out good people. The genetic factor will load the gun and lifestyle triggers the bullet. And hence happiness and peace vanished from our life and become a regular visitor of hospital. Because of this phenomenon all hospitals in Kerala and other places are full, no room or bed in hospital, car parking always full, waiting time to see the doctor is increased, Very high consultation fees and expensive drugs has to swallow for longer period are the spin off. This proves that money alone cannot buy good health and happiness in human life. An attitude change is also needed for a quality life with our presence how much happiness we can give to others will make you productive and happy.

As responsibilities increases the ability to manage stress also increasing. That is why the American president, prime minister of India, chief ministers of states,

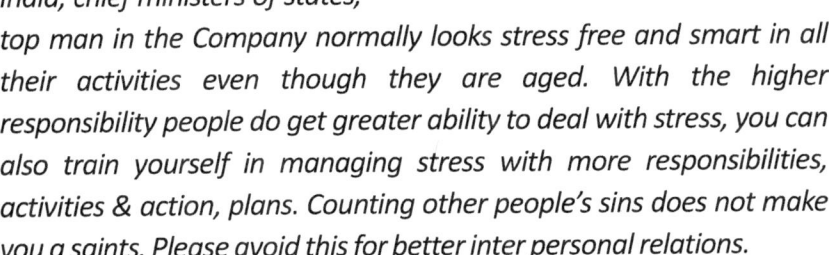

top man in the Company normally looks stress free and smart in all their activities even though they are aged. With the higher responsibility people do get greater ability to deal with stress, you can also train yourself in managing stress with more responsibilities, activities & action, plans. Counting other people's sins does not make you a saints. Please avoid this for better inter personal relations.

The second aspect to reduce stress is to identify the source of stress. There are basically three sources of stress. As we told in the case of Gulf countries & Techno park the first is "job ambiguity".

When you take up a job that is not well studied & structured, there can be a lot of confusion about your role and this can create a stress in most people. Solution for this is to go for training and develop a passion to that job or try to get another  job you love. Introduction of new Apps like internet banking mobile phone etc can reduce stress related to memory loss. Old people must write in a book all fixed deposit details of them.

The third source is "relationship stress", people are not having good relation with the team in office or family members in the house. Most of the marriages now fail because of relationship stress. Culprit behind the divorcee is expectations, which reduces the joy in the family. Husband, wife, in laws will expect more wealth and luxury, forgetting to love each other. Family members love gold ornaments, flat and house, costly car and speaking only to mobile phone. Not smiling each other will create a stressful environment in the house and this will end up in relationship stress and communication brakes forever. Usually the problem between husband and wife is that a wife wants assurance from her husband that she loved by him. Husband will think that he loves his wife, does not know how to show this to her. A man is not always expressive like women. Woman demands an expression of love, which a man is unable to do. He cannot tell the wife how much he loves. Only few couple knows LOVE is Limitless Oscillating Vibrating Energy. This is a clean energy to be generated within the family and sheared each other unconditionally. If you start judging others, you do not have time to love them. One day you also to be judged by others and that will be end of the news. House is *made with bricks; home is made with love, sympathy, empathy, concern, self respect etc. What you sow only you reap. What you give only you get. What you see in others also exists in you. You are the makers of your life miserable. You make yourself the looser. You make good and evil. It is you who put your hands before your own eyes and say the world is dark. Today is the best day to take your hands away and see the light and enjoy the stress free life. Human mind is like a parachute, it works better when it is fully open. Some long term tips are given below to open your parachute if it is stuck.*

In case of stress created through relationship problem or in –law problem there is a five-step long term process to reduce stress. It is experiment able and experience able.

Step 1 – WAIT - *with time the relationship can improve and that can reduce stress .Apply the effect of forgetfulness and forgiveness, The two gifts of beautiful mind are forgetfulness and forgiveness. It is easy to implement and it is non destructive. Give importance to relations rather than the ego. Be a self esteem person. With control over the tongue it is very easy to talk the small mistake of others it is very difficult to explain our own big mistakes. Also see that the problem makers take advantage of this attitude sometimes to trigger you. So waits for your turn to come ("Every dog has a Day").*

Step 2 – NEGOTIATE - *if wait fails, one can sit across the table or in a family get together, and one can negotiate as to what can be done in the better relationship in the future. Explain clearly what you deserve, never negotiate for your desire. Most of the in-law stress and possessive stress can be settled through negotiation within the family. The happiest of people what we come across do not necessarily have the best of everything. They just make the most of everything that comes along their way by negotiation.*

Step 3 – FIGHT - *if negotiation fails you tell the person that the* relation ship problem is very serious and you cannot tolerate much longer. The fight can end up either you win or you lose or file a case in the vanitha police station and struggle. Be bold when you lose, be soft when you win. People will appreciate when you are in bright light. Every one likes to fallow your path. When you are in dark even your own shadow does not follow you be care full while fighting.

Step 4 – FLIGHT - if fight fails, you should be prepared to go away and find another job and another house to live. After all our precious life is too short to quarrel and fight each other for protecting the ego.

Step 5 – SURRENDER - if nobody is giving a job or not able to find a new alliance, a house to live, you accept the stress. Accept the relationship problem as a part of our life and get tuned with it and change our attitude instead of changing the attitude of others. Remember, attitude shows the altitude. Surrendering for an honest relation is good because honest relations are just like drinking water no color, no shape, no taste to explain but still very important for our life. No one can go back and change the wrong start but any one can start now and create a successful ending for a better life.

Step 6 – CURSE - if all other step fails do the most popular layman approach to deal the stress, is to curse and believe in fate. It does not improve relations; hence this is the step that we have to avoid. Then go from 1 to 5 (wait – negotiate - fight – flight - surrender) this is the way we can deal with the problem of stress created through relationship in Office, family & in -law problems. Mistakes are painful. But a collection of mistakes lead to experience and make you successful. The most beautiful thing in this world is to see your parents smiling, knowing that you are the reason behind that smile. Please do not curse for what tragedy happened you, because parents are the only true well wishers of you in the world.

Apart from the long-term plan (wait, negotiate, fight, flight,

surrender) for reducing stress. I would like to suggest three more short-term plans.

Plan 1 – PRAYING to GOD, prayer is not a spare wheel that you pull out when you are facing problems, but it is a steering wheel that directs the right path throughout our life.

Plan 2-YOGA & MEDITATION - do 'pranayama' & 20 Soorya Namaskaram daily, deep breathing is very useful to bring down the blood pressure. Watching the AASTHA the faith TV channel morning 5 am to 8 am evening 8 pm to 9.30 pm for yoga and meditation practice daily. Avoid seeing TV serials during this time. When you feel that you are under stress take 2 to 3 glass of warm water. Do the deep breathing. If the stress is not reducing then wash your face with cold water & squash your hands 10 to 20 times by holding a handkerchief. Or you can use a rubber ball. Practice meditation daily 30 mts. Eat more vegetables than meats, drink more water.

Plan 3 – Identifying a DOCTOR or a GOOD FRIEND or a MENTOR and tell about your stress problems or call him/her through phone and vent your stress. This mentor can be a college or a family member. Extreme care has to be taken for the selection of the mentor, since you are going to tell your secrets to him or her must be a well-wisher of you and your family.

Apart from these relationship and job ambiguity problems, the third source of stress is "personal problems". After all, a person does not live separately. Eight hours a day in the work place and 16 hours a day out of the work place. A person lives 24 hours a day continuously. What happens in one location has an impact on what happens to him at the other location. The stresses become accumulated. If personal problem is mental then it is a serious issue.

SPECIAL STRESS PROBLEM.

However, there are two special stress problems, which we must look in to. These are Economic stress & Social stress. These two are complementing each other. These two stresses are invited stress due to lack of knowledge about our NEED & WANT. We need a car, but we want three cars each costing more than ten lakhs, similarly people purchase more house and flats. We may be cheated by money chain, on line purchase, giving the details of ATM cards to others , wrong usage of inter nets, surety for chitty, lending money , borrowing money etc. This will end up in economic & social stress.

When we reaches about 60 years of age, suddenly experiences a great deal of frustration. If the person is reasonably intelligent, from the time he starts his job till the age of 55, he can easily get a good bank balance and is quite happy about this, but when he looks ahead from the retirement day, he finds that till his death there may be no more incomes. In addition to this expected retirement, he gets frustration due to health issues.

This is the time when the children start frustrating the parents by divorce for simple reasons, flat booking, demanding child care, compelling to live at an extreme climate place, hospital test, treatment with costly medicines. They do not sure the medicines are really required etc The frustration also adds up and so this becomes difficult period for a retirement life. Old people also have the right to continue in the same status with own house, car, good food etc till their death. So keep adequate bank balance always. If you ask to your children they may not be a position to help you when you need the money. In some families children will comment on the luxury of the old people spending their own time and hard earned money and advising to invest in wrong places. Please do not accept all these comments it will only give you more stress. Some people go for new jobs and getting in to problems after age 60. Memory and vision problems will also create stress. No solution for getting old is the reality, accept it.

To reduce this stress, we must think of developing on "alternate focus in life". A person not getting happiness from his family gets very great distress for a long period. The average mortality rate in Kerala is now gone up to 90 years. But if he has some other activity like writing, gardening, swimming, social club, professional association, residents association -etc. then he finds that his frustration in one area is compensated by success in another. For this activity good health is also required in the old age. The best retirement benefits are physical health, mental health, family health, economical health, social health and spiritual health. Try to avoid power struggle & worrying about what happened. It does not take away tomorrow's troubles, but it takes away today's peace.

All things in life are temporary. If going well enjoys it, they will not last forever. If going wrong do not worry, they cannot last longer either. Nothing is permanent in life. You can win life by all means. Simply avoid two things, comparing and expectations both will reduce joy. Wish you good health.

ആഗോള താപനവും, കാലാവസ്ഥാ വ്യതിയാനവും" തൻമൂലമുള്ള വെല്ലുവിളികളും
...ഒരവലോകനം

ഇനി വരുന്ന തലമുറകൾക്ക്, ഈ ഭൂമിവാസയോഗ്യമാണോ? അല്ല കാരണം മലിനമായി വറ്റി വരണ്ട പുഴകളും നദികളും, ഡാമുകളും, കിണറുകളും ഉണങ്ങി വരണ്ട വയലേലകളും, മലീനസമായ പ്രാണവായുവും, മായം കലർന്ന ഭക്ഷണവും, മരുന്നും, ചിലവേറിയ ചികിൽസയും അങ്ങിനെ ധാരാളം അധർമ്മങ്ങളാണ് അവരെ കാത്തിരിക്കുന്നത്. ദിവസവും പകൽ, രാത്രി, ഇവ മാറി വരും. ഉദയമില്ലാതില്ലാ അസ്തമയം, ഉണരു മനസ്സേ ഉണരു. പ്രകൃതിയിൽ സംഭവിക്കുന്ന എല്ലാത്തിനും ഒരു കാരണം ഉണ്ടെന്നാണ് മേൽ വരികൾ സൂചിപ്പിക്കുന്നത്. അതേ സമയം ഇനി വരും തലമുറക്കായി ഭൂമിയെ രക്ഷിക്കുന്നതിന് മനുഷ്യമനസിനോട് ഉണർന്നു പ്രവർത്തിക്കാനും പറയുന്നു.

ആഗോള താപനത്തിനും, കാലവസ്ഥാ വ്യതിയാനത്തിനും കാരണം മനുഷ്യമനസ്സിന്റെ ഭൗതിക നേട്ടത്തിനായി ഉള്ള അമിത പ്രകൃതി ചൂക്ഷണവും മലിനികരണവും ആണ്. സൂര്യപ്രകാശത്തിൽ അടങ്ങിയിട്ടുള്ള അൾട്രാ വയലറ്റ്, ഇൻഫ്രാ റെഡ് തുടങ്ങിയ തരംഗദൈർഘ്യം കുടുതലുള്ള എല്ലാ രശ്മികളും മനുഷ്യനും, ജീവജാലങ്ങൾക്കും ആപത്തായതിനാൽ ഈശ്വരൻ അന്തരീക്ഷത്തിൽ ഒസോൺ പാളികൾ നൽകി രക്ഷിച്ചു. അതിന് വിള്ളലുകൾ ഉണ്ടാക്കാൻ

Infinite Space & Unlimited Excitement

മനുഷ്യൻ അമിത അളവിൽ CO_2, കാർബൺഡൈഓക്സൈഡ്, CH_4 മിഥേൻ (Methane) നീരാവി ഇവ ഉല്പാദിപ്പിച്ചു. ഡീസൽ, പെട്രോൾ ഇവയുടെ അമിത ഉപയോഗവും, സിമന്റ്, കൽക്കരി, പ്ലാസ്റ്റിക്ക് ഇവയുടെ അമിത ഉപയോഗവും, പ്രകൃതിയെ മലിനമാക്കി. കാട് നശിപ്പിച്ചു! മലകൾ ഇടിച്ച് നിരപ്പാക്കി, ജലാശയങ്ങൾ മലിനമാക്കി, വായു മലിനമാക്കി അനേകം നദികൾ, കാടുകൾ. കിണറുകൾ, കടൽ, കായൽ മത്സ്യങ്ങൾ, ചിത്രശലഭം എന്നു വേണ്ട പലതും ഈ ലോകത്തു നിന്ന് അപ്രത്യക്ഷമാവുമ്പോഴേ മനുഷ്യൻ ചിന്തിക്കൂ ഇവ എന്താണ് നമുക്ക് നൽകിയിരുന്നത് എന്ന്. അപ്പോഴേക്കും കാര്യങ്ങൾ പിടിവിട്ട് പോയിരിക്കും. വംശനാശം സംഭവിച്ചിരിക്കും. ഇങ്ങനെ മലിനീകരണത്തിന്റെ അളവ് കൂടുന്നതിനാൽ അന്തരീക്ഷത്തിന്റെ ഏറ്റവും മുകളിലത്തെ ഓസോൺ പാളിയിലുള്ള വിള്ളലുകൾ കൂടിവരുന്നു. അതനുസരിച്ച് സൂര്യൻ ഭൂമിയുടെ പ്രതലത്തെ ചൂട്ട് പഴുപ്പിക്കുന്നു. ഈ താപം റേഡിയേഷൻ മൂലം പുറത്തുപോകാൻ ടൺ കണക്കിനുള്ള കാർബൺഡൈഓക്സൈഡ്, മിഥേൻ (Methane), നീരാവി ഇവ അനുവദിക്കാതെ വരുന്നു. ഈ പ്രതിഭാസത്തെ ശാസ്ത്രം, *"green house effect"* എന്ന് വിളിക്കുന്നു. മിഥേൻ, കാർബൺഡൈഓക്സൈഡ് ഹരിതവാതകങ്ങളാൽ ഭൂമി ഏതാണ്ട് വെയിലത്ത് ഗ്ലാസെല്ലാം അടച്ചുക്കിടക്കുന്ന ഒരു കാർ മാതിരി അകം ചൂടായി ഇരിക്കുന്നഅവസ്ഥയിലാണ്. നാൾക്കുനാൾ ഭൂമിയുടെ താപനില ഉയർന്നു വരുന്ന ഈ പ്രതിഭാസത്തെയാണ് **"ആഗോളതാപനം"** എന്ന് പറയുന്നത്.

ശാസ്ത്ര ലോകം ഇതിനെ ഒറ്റവാക്കിൽ ഒതുക്കി *Catastrophic"* അതായത് **"ഭയാനകം"** ഇന്ത്യ, പ്രത്യേകിച്ചും കേരളത്തിന് കൂടുതലും തീരദേശം

ഉള്ളതിനാൽ 2050-2100-ാം ആണ്ടോടുകൂടി സമുദ്രനിരപ്പ് ശരാശരി അഞ്ച് അടി ഉയരുമെന്ന് കണക്കാക്കപ്പെടുന്നു. ഭൂമിയുടെ രണ്ടു ധ്രുവങ്ങളിലും ഉള്ള മഞ്ഞുമലകൾ ചൂടിൽ ഉരുകി, വെള്ളമായി കടലിൽ വരുന്നതിനാലാണ് ഇത് സംഭവിക്കുന്നത്. കൂടാതെ മഴയും, വെള്ളപൊക്കവും പ്രവചനാതീതമാകുമെന്നും, കൊടുങ്കാറ്റും, ചുഴലിക്കാറ്റും ഉണ്ടാവുന്ന ഇടവേള കുറഞ്ഞുകാണരുതെന്നും മുന്നറിയിപ്പ് നൽകുന്നു. ആകാശം ഇടിഞ്ഞുവീഴും എന്നൊക്കെ പണ്ട് പറയാറുള്ളത് മേഘവിസ്ഫോടനം എന്ന പുതിയനാമത്തിൽ *"Cloud burst"* ആയി മഴപെയ്യും? ഒരു ചെറിയ പ്രദേശത്ത് കാർമേഘങ്ങൾ ഉരുണ്ടുകൂടി അവിടെത്തന്നെ അതിശക്തമായ മഴ മണിക്കൂറിൽ പത്തുസെന്റീമീറ്റർ കൂടുതൽ മഴ ചെന്നയിലും, ഉത്തരാഘണ്ഡിലും, കാശ്മീരിലും നമ്മൾ കണ്ടതാണ്. ഇത് ഇനിയും ആവർത്തിക്കും കേരളം ഉൾപ്പെടെ പല സ്ഥലങ്ങളിൽ പല തവണ പേമാരി പെയ്യും. തന്മൂലം അങ്ങനെ പ്രവചിക്കാൻ പറ്റാത്ത കാലാവസ്ഥ ഉണ്ടാവുമ്പോൾ വിതക്കുമ്പോൾ പെയ്യേണ്ട മഴ (കാലവർഷവും, ഇടവപ്പാതിയും, തുലാവർഷവും) കൊയ്യാറാവുമ്പോൾ പെയ്യും. തന്മൂലം കൃഷിനാശം, കുടി വെള്ളക്ഷാമം. ഇവ വരും കാലങ്ങളിൽ പ്രതീക്ഷിക്കാം. ഈ

വർഷം 2017 തന്നെ ഇതിന് ഉദാഹരണമാണ്, ഇത് മഴ കുറഞ്ഞ വർഷം, ജലക്ഷാമം രൂക്ഷം ഇതിനാണ് *"കാലാവസ്ഥ വ്യതിയാനം"* എന്ന് പറയുന്നത്. ലോകത്തിന്റെ പലഭാഗത്തും കാലാവസ്ഥ കാലം തെറ്റി വന്നുകൊണ്ടിരിക്കുന്നു. പ്രവചിക്കാൻ പറ്റാതെ UAE, DUBAI ലുള്ള മരുഭൂമിയിൽ വരെ കനത്ത മഴ പെയ്യുന്നു. സമുദ്ര നിരപ്പ് ഉയരുന്നത് കേരളത്തിൽ പ്രത്യേകിച്ചും, ഇന്ത്യയിൽ കൂടുതലും പ്രശ്നങ്ങൾ ഉണ്ടാകും. കാരണം ഇവിടെ തീരദേശം മറ്റു രാജ്യങ്ങളെക്കാൾ കൂടുതലാണ്. ഇതിനാലാണ് ഈ സ്ഥലങ്ങളിൽ പുതിയ നിർമ്മാണ പ്രവർത്തനങ്ങൾക്ക് വിലക്ക് ഏർപ്പെടുത്തിയിരിക്കുന്നത്. സമുദ്ര നിരപ്പിൽ ചൂട് അധികമായിരിക്കും, അതിനാൽ ബാഷ്പീകരണം കൂടും. ഉയരം കൂടുതോറും തണുപ്പും കൂടും. ഊട്ടി, കൊടൈക്കനാൽ ഇവ സമുദ്രനിരപ്പിൽ നിന്ന് ഉയർന്നാണ് അതിനാൽ തണുപ്പ് അനുഭവപ്പെടുന്നത്.

എല്ലാ രാജ്യത്തും ചൂട്, തണുപ്പ്, Humidity ഇവ ദിവസവും

അളക്കുന്നുണ്ട്. ഇവിടെ തിരുവനന്തപുരത്ത് മ്യൂസിയത്തും, വിമാനത്താവളത്തിലും, താപനില രേഖപ്പെടുത്തുന്നുണ്ട്. കുറേവർഷങ്ങളായി ഇത് രണ്ട് ഡിഗ്രിയിൽ കൂടുതൽ തിരുവനന്തപുരത്ത് തന്നെ കൂടുന്നതായി കാണുന്നു. മറ്റുപല രാജ്യങ്ങളിലും താപം ഉയർന്നാണിരിക്കുന്നത്. ഇതിനെയാണ് **ഗ്ലോബൽ വാമിങ്ങ്** അല്ലെങ്കിൽ **ആഗോള താപനം** എന്ന് പറയുന്നത്. ഇങ്ങനെ പോയാൽ അടുത്ത പത്തുവർഷം

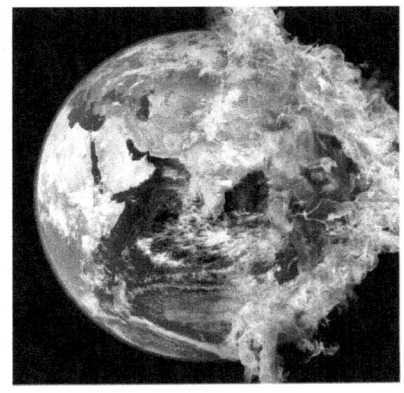

2027 ൽ ഇത് അഞ്ച് ഡിഗ്രി വരെയാകാം. (പകൽ താപനില 50 ഡിഗ്രിയിൽ കൂടുതൽ.) ഇത് പലവിധത്തിൽ സാധാരണ ജീവിതത്തെ ബാധിക്കും.

ലോക ജനത ചിലപ്പോൾ പരസ്പര സഹകരണമില്ലാതെയും, സ്നേഹമില്ലാതെയും, ചപ്പുചവറുകളും, പ്ലാസ്റ്റിക്കും ദിവസവും കത്തിച്ച് കുറേക്കാലം ജീവിച്ചേക്കാം പക്ഷെ, കുടിവെള്ളം, പ്രാണവായു ഇവ ഇല്ലാതെ എത്ര നാൾ ജീവിക്കും? ആഗോള താപനവും, കാലാവസ്ഥാ വ്യതിയാനവും കൂടി വലിയ തോതിൽ ജലക്ഷാമം ഉണ്ടാകും. ഒരു കാലത്ത് സൗജന്യമായി ലഭിച്ചിരുന്ന കുടിവെള്ളം ഇന്ന് കോടികളുടെ കച്ചവട വസ്തുവാണ്. ഭാവിയിൽ പ്രാണവായുവിനും വില കൊടുക്കേണ്ടിവരും. ഇപ്പോൾ തന്നെ അരിവില 50 രൂപയിൽ കൂടുതലാണ്. പച്ചക്കറിക്കും മീനിനും തീ വിലയാണ്. ഇവ ഭക്ഷ്യ ക്ഷാമത്തിലേക്ക് വഴിതെളിക്കും. പകൽ സമയം പാടത്തും, കടലിലും പണിയെടുക്കാൻ പറ്റാതെ കാർഷികമേഖല, തീരദേശ മേഖല ഇവ കൃഷി നശിച്ചും, മത്സ്യം കിട്ടാതെയും വലയും. ഇതെല്ലാം ഭയപ്പെടുത്താൻ എഴുതുന്നതല്ല. അടുത്ത തലമുറ കരുതിയിരിക്കാൻ വേണ്ടി എഴുതുന്നതാണ്. കടലിൽ നിന്നും കുടിവെള്ളം ഉല്പാദിപ്പിക്കാൻ കഴിയണം. ഡാമിൽ വെള്ളം ഇല്ലാത്തതിനാൽ വൈദ്യൂതി ക്ഷാമം വരും അതിനാൽ സൗരോർജ്ജം പരമാവധി ഉപയോഗിക്കണം. ഉള്ളതിനെ സൂക്ഷിച്ച് ഉപയോഗിക്കണം.

2017 ൽ ലോകജനത 740 കോടി എന്ന് കണക്കാക്കപ്പെടുന്നു. ഇത്രയും ജനതക്കും ഇപ്പോൾ തന്നെ എല്ലാം സുലഭമല്ല. 2053 ൽ ലോകജനത 1000 കോടി കവിയും അപ്പോഴത്തെ സ്ഥിതി വളരെ ഗുരുതരമായിരിക്കും.(*Water every where but a single drop is not available for drinking*)

ഇപ്പോഴത്തെ അവസ്ഥയിൽ കുറേകാലം കഴിയുമ്പോൾ അതായത് 2050-2100 കാലത്ത് സമുദ്രനിരപ്പിലോ അതിൽ താഴെയുള്ള പ്രദേശത്തോ വലിയ ദുർഗതി തന്നെയായിരിക്കും. ആഗോള താപനം കണ്ടു വെച്ചിരിക്കുന്നത്. കായലും, വയലുകളും കൈയ്യേറി നിരത്തുകളും പാർക്കുകളും ടൈൽസ് പാകി മഴവെള്ളം മുഴുവനും കടലിൽ ഒഴുക്കി കളയുന്ന പുതിയ ജീവിതരീതി കാരണം കുടിവെള്ള ക്ഷാമവും കൃഷിനാശവും പുതിയ തലമുറക്ക് ഈ ഭൂമി വാസയോഗ്യമല്ലാതായി തീർന്നേക്കാം. അതിനാലാണ് NASA, ISRO തുടങ്ങിയ സ്ഥാപനങ്ങൾ ചൊവ്വയിൽ വെള്ളം ധാതുക്കൾ ഇവക്കായിപര്യവേഷണ പദ്ധതികൾ രൂപകൽപന ചെയ്യുന്നത്.

ഭൂമിയുടെ ചുറ്റുമുള്ള ഓസോൺ പുതപ്പിലെ വിള്ളലുകളിലൂടെ അരിച്ചിറങ്ങുന്ന സോളാർ റേഡിയേഷൻ, ഇൻഫ്രാ റെഡ്,രശ്മികളായി പരിണമിച്ച് അവ ഓസോൺ, (O_3) മീതേൻ (CH_4), നീരാവി, കാർബൺഡൈ ഓക്സൈഡ് (Co_2) എന്നിവയിൽ ലയിച്ച് ക്രമേണ താപനില ദിനം പ്രതി ഉയർത്തുന്ന പ്രവണതയെ *"ഗ്രീൻ ഹൗസ് ഇഫ്ക്ട്"* എന്ന് പറഞ്ഞുവല്ലോ? ഇതിന് നാം ഓരോത്തരും ഉത്തരവാദികളാണ്. ഇത്രയൊക്കെ ഇതിന്റെ കാരണങ്ങൾ കണ്ടുപിടിച്ചിട്ടും, മനസ്സിലാക്കിയെങ്കിലും ഇതിനൊരു പരിഹാരം വളരെ ഒച്ചിഴയുന്ന വേഗതയിലാണ് നടക്കുന്നത്.

ഇന്ത്യയിൽ വാഹനങ്ങൾക്ക് പുതിയ മലിനീകരണ നിയന്ത്രണവും, കൂടുതൽ മലിനീകരണം ഇല്ലാത്ത പെട്രോൾ, ഡീസൽ, വായു മലിനീകരണത്തിന്റെ തോത്കുറക്കാൻ സഹായിച്ചേക്കാം. കൂടാതെ കൃത്രിമമഴ മഴപെയ്യിക്കുവാൻ കേരളത്തിൽ ചില നീക്കങ്ങൾ നടക്കുന്നുണ്ട്. "Cloud Seeding" എന്ന പ്രവർത്തി വളരെ ചിലവേറിയ പദ്ധതിയായതിനാൽ ലഭ്യമാകുന്ന സാധാരണ മഴ വെള്ളം ഭൂമിയിൽ താഴാനുള്ള പദ്ധതിക്ക് മുൻഗണന കൊടുക്കുന്നതാണ് കൂടുതൽ പ്രായോഗികത എന്നാണ് ശാസ്ത്രലോകം വിലയിരുത്തുന്നത്. ലോകരാഷ്ട്രങ്ങൾ G20, UN പലപരിപാടികളും പ്ലാൻ ചെയ്തിട്ടുണ്ട്. ഒന്നിനും ഈ വർദ്ധിച്ചുവരുന്ന ചൂടിന് ഒരു പരിഹാരം കണ്ടെത്താൻ കഴിഞ്ഞിട്ടില്ല. ആഗോള താപം രണ്ട് ഡിഗ്രിയിൽ താഴെ കുറക്കാൻ കോടികളുടെ ചിലവാണ് ഇതിന് തടസമാകുന്നത്. ഇത് ഒരു ആഗോള പ്രതിഭാസമായതിനാൽ എല്ലാ രാജ്യങ്ങളും വാഹനങ്ങളിലെ കാർബൺഡൈഓകാസൈഡ് പുറം തള്ളുന്നത് നിയന്ത്രിക്കാൻ രൂപകല്പനയിൽ മാറ്റം വരുത്തി, വൈദ്യുതി ഉപയോഗിച്ച് ഓടിക്കാൻ പറ്റിയ ഡിസൈനുകൾ വാഹനങ്ങളിൽ നടപ്പിലാക്കണം. കൽക്കരിയും, ഡീസലും കത്തിച്ച്വൈദ്യുതി ഉല്പാദനം നിർത്തിപകരം കാറ്റാടി,

സോളാർ, ന്യൂക്ലിയർ വൈദ്യുതി നിലയങ്ങൾ സ്ഥാപിക്കുക, അന്തരീക്ഷ മലീനികരണം നടത്തുന്ന വ്യോമയാന കമ്പനികൾക്ക്, ഫാക്ടറികൾക്ക്, വലിയ പിഴ ഈടാക്കുക, ഇപ്പോഴുള്ള കാർബൺ ക്രെഡിറ്റ് സംവിധാനം വിപുലമാക്കുക, വീടുകളിൽ പ്രഷർ കുക്കർ, മാലിന്യങ്ങളിൽ നിന്നും കുക്കിംഗ് ഗ്യാസ്. പുതിയ ചവർ സംസ്ക്കരണ സംവിധാനം കുടുംബശ്രീ, റസിഡൻസ് അസോസിയേഷൻ, കോർപ്പറേഷൻ സംയുക്തമായി നടത്തുക, കൂടുതൽ കിണറുകൾ, മഴവെള്ള സംഭരണികൾ ഇവയിൽ മഴ വെള്ളം പിടിക്കുക.

നമ്മുടെ ഈ ഭൂമിയെ അടുത്ത തലമുറക്ക് ജീവിത യോഗ്യമായി ഏൽപ്പിക്കേണ്ട ഉത്തരവാദിത്വം, സന്തോഷത്തോടെ ഏറ്റെടുത്ത് നാം ഓരോത്തരും നമ്മളാൽ കഴിയും വിധം "അണ്ണാറക്കണ്ണനും തന്നാലായത്" എന്നു പറയുന്നതുപോലെ പലതും പ്രവർത്തിച്ചേ മതിയാകൂ. എല്ലാ ഭരിക്കുന്ന സർക്കാരോ, യുണൈറ്റഡ് നേഷനോ ചെയ്യുമെന്ന് വിചാരിച്ചു ഉത്തരവാദിത്വത്തിൽ നിന്നും ഒഴിഞ്ഞുമാറാതെ നമുക്ക് എന്തുചെയ്യാൻ പറ്റുമെന്ന് ചിന്തിക്കാനും കൂട്ടായി പരിശ്രമിക്കുവാനും ഈ ചെറുലേഖനം സഹായിക്കും

"ലോകാ സമസ്താ സുഖിനോ ഭവന്തു'; ശുചിത്വം വീടിനും നാടിനും അത്യാവശ്യമാണ്. ഈ ഭൂമിയെ സംരക്ഷിക്കുവാൻ സ്വയം തീരുമാനം എടുക്കുവാൻ അപേക്ഷിച്ചുകൊണ്ട് നിർത്തുന്നു. നൻമകൾ നേരുന്നു.

www.ingramcontent.com/pod-product-compliance
Lightning Source LLC
Chambersburg PA
CBHW052331220526
45472CB00001B/373